Caroline Williams is a science journalist and editor. A feature editor and regular contributor to *New Scientist*, her written work has also appeared in *The Guardian*, BBC Future, and BBC Earth, among others. She has worked as a radio producer and reporter for BBC Radio, and was the regular co-host of the *New Scientist* podcast from 2006 to 2010. She is editor of the *New Scientist* Instant Expert guide *How Your Brain Works: inside the most complicated object in the universe* (John Murray, 2017). She holds a BSc in biological sciences from Exeter University and an MSc (Distinction) in science communication from Imperial College London. She lives in Surrey, UK.

*For Jon and Sam, with love.*

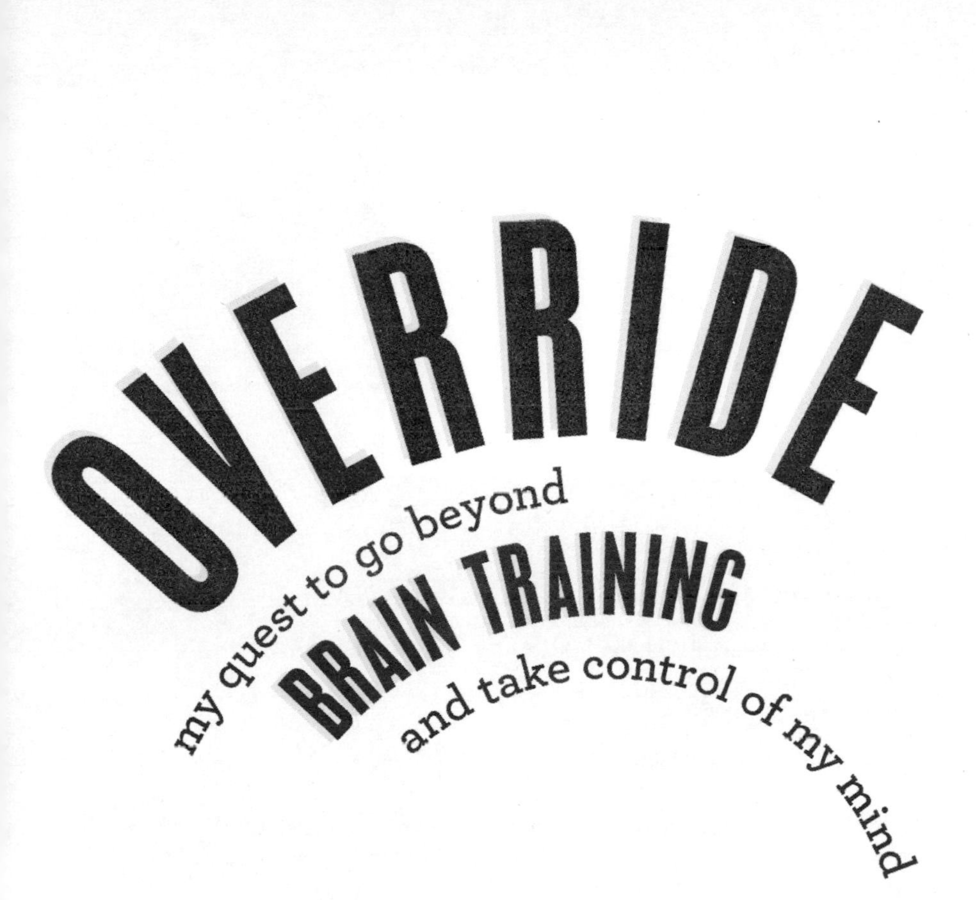


CAROLINE WILLIAMS

SCRIBE
Melbourne • London

Scribe Publications
18–20 Edward St, Brunswick, Victoria 3056, Australia
2 John Street, Clerkenwell, London, WC1N 2ES, United Kingdom

First published by Scribe 2017



Typeset in 11.5/17 pt Scala by the publishers
Printed and bound in the UK by CPI Group (UK) Ltd, Croydon, CR0 4YY

Scribe Publications is committed to the sustainable use of natural resources and the use of paper products made responsibly from those resources.

9781925321906 (ANZ edition)
9781925228984 (UK edition)
9781925307931 (e-book)

A CiP record for this title is available from the National Library of Australia and the British Library.

scribepublications.com.au
scribepublications.co.uk

# Contents

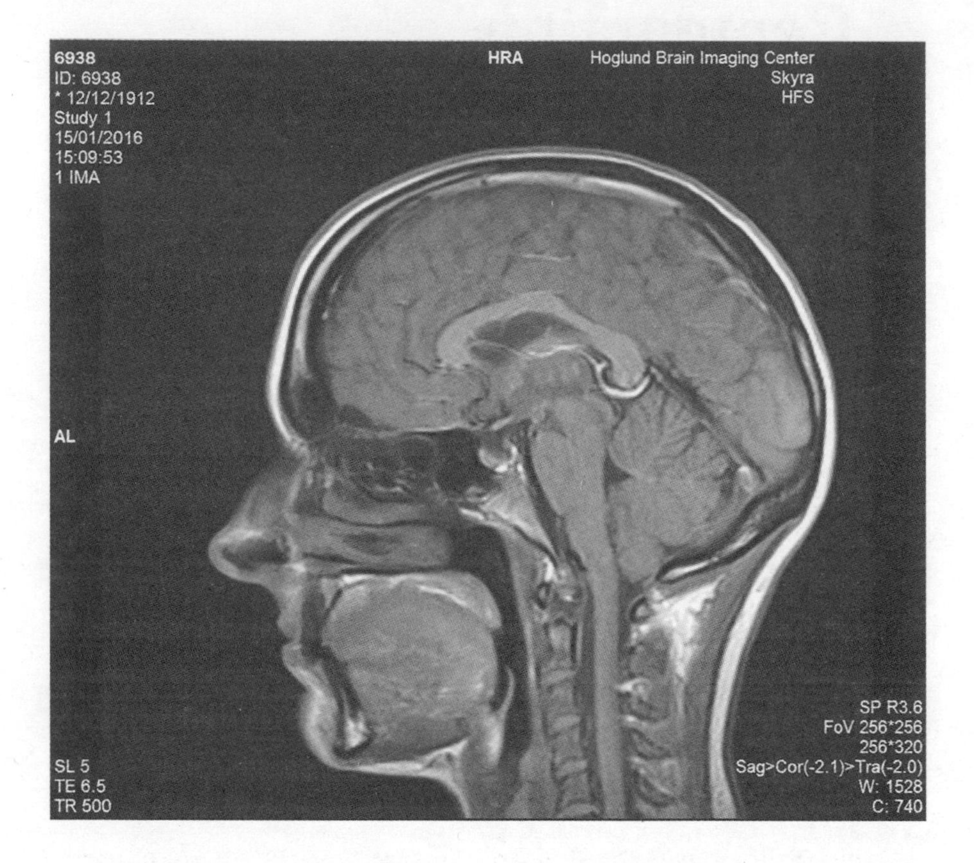


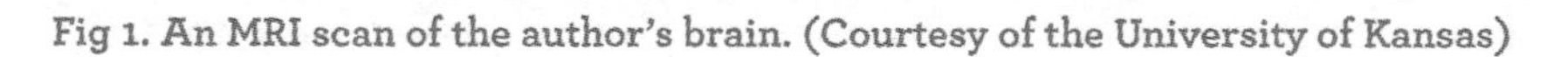

Fig 1. An MRI scan of the author's brain. (Courtesy of the University of Kansas)

# Introduction

Heathrow airport is huge. So if you happen to leave your hand luggage in the departure lounge, and don't notice until you are at gate 21a and the flight is boarding, it's quite a long way to run back — about a 15-minute round-trip, in fact, although it feels a lot longer when you have just heard a stern announcement that 'luggage left unattended will be removed and may be destroyed'.

Thankfully, my bag was in one piece, and exactly where I left it. I found it in a shop just as the assistant was about to call security, stammered an apology through a dry mouth, and pelted back to the gate in time to catch my flight. It wasn't until I'd calmed down, with a stiff gin and tonic in hand, that I realised this sort of mishap was exactly why I was taking this flight in the first place.

I was on my way to Boston, Massachusetts, to meet two neuroscientists who carry out research into sustained focus and attention. My hope was that they would help me find a way to override my natural tendency to be stressed and distracted; to help me replace it with a calm focus that I could sustain for a useful amount of time. It was the first step in a journey that was to last more than a year, and take me back and forth across the US and Europe in search of real-life fixes for my brain's shortcomings.

I wanted to apply the best that modern brain science had to

offer, and to get a glimpse into the future of real-life brain training. Focus was just the beginning. In the months that followed, I would try science-based interventions for, among other things, a non-existent sense of direction, an unhealthy worrying habit, and some embarrassingly bad number skills. Then I would branch out into some more mysterious corners of the mind, such as creativity and the perception of time.

There are good reasons to think that mine was a journey worth taking. First, there is a decade's worth of evidence that the brain is 'plastic' — it retains the ability to change physically in response to the things we learn and experience throughout life. As a science journalist and former feature editor at *New Scientist* magazine, I have, over the years, written tens of thousands of words on so-called neuroplasticity, and as time went on, I became more and more curious about how I might apply this to my own brain.

But when I started looking for answers, what I found was ... nothing, of any real, practical use. Despite all of the research into the brain's awesome powers of plasticity, no one seemed to know exactly what we should be doing to apply the science to everyday life. Sure, there are fascinating tales of people harnessing their brain's plasticity to recover from major brain injuries, but to my knowledge there was no such evidence for the average person on the street.

To me, it seemed like a pretty major hole in the assumption that neuroplasticity can be applied by anyone. For a start, injured brains are very different to healthy ones. After a stroke, the brain releases various growth-promoting chemicals at the site of injury to try and repair the damage. The same degree of 'rewiring' may not be possible when there isn't a major roadblock that the brain needs to work around. On the other hand, it's hardly surprising that we can learn new skills throughout our whole lives — learning and remembering are what brains are designed to do.

With 86 billion neurons, and trillions of connections, an adult brain is a pretty remarkable feat of engineering. By the time it gets to maturity, it has already been on an incredible journey. A large part of the job of the adult brain is to work as a kind of pattern-spotting and generalisation machine — whirring away in the background, making sense of what is happening now and how it relates to what has already been stored in memory.

These memories can only come from experience, which is why babies and children come primed to learn, with an endless supply of curiosity about what things are like and why. Once this groundwork is complete, a lot of the day-to-day processing of the brain is done on autopilot, with unconscious processing taking care of working out what is happening and how we should respond. The brain does this for a good reason: unconscious processing is fast and effortless, and leaves plenty of thinking-power free to concentrate on things that need more focus.

The learning process starts surprisingly early: in the last few weeks before birth, a baby's brain is hard at work, forming strong memories of its mother's voice and the sounds of the world it will be born into. It also learns from its mother's physical state — a high dose of stress hormones from the mother, for example, programmes a baby's brain to develop in ways that leave it more reactive to stress in later life. The brain is learning that it needs to be on alert because it is being born into a dangerous world.

In so many ways, what we experience in early life shapes the adult that we will become, deciding which assumptions our brains will make without bothering to inform the consciousness. This, combined with each person's particular genetic inheritance, means that each brain in adulthood is not only totally unique, it got that way more or less by accident: the outcome of a genetic and life-experience lottery.

If neuroplasticity can be applied in adulthood, though, it provides an opportunity to change all that; to take a fresh look at the brain you have ended up with as an adult and decide what to keep and what to change.

There is only one thing that worries me about this process — and it's something that I admit I hadn't thought of until I was enthusiastically telling a friend about my plans. His reaction wasn't at all what I expected. Stephen, a friend from my yoga class, looked horrified at the very *idea* of trying to change your brain. 'But surely you are a unique and wonderful person who isn't like anyone else,' he said. 'Why would you want to change that?' It threw me for a while, because it's true that my brain, warts and all, is the very thing that makes me *me*. If I change it, there is always the risk that I won't still be me afterwards.

On the other hand, if brain change continues throughout life, then the work of making me *me* isn't yet done. Why live with the less than helpful bits if the wonders of neuroplasticity mean that you don't have to? Most of us don't even decide what we want to do with our lives until long after our brains have become stuck in their ways. As a result, we adults we spend much of our time drifting along in the passenger seat of our own minds. Wouldn't it be nice to jump into the driver's seat for a change?

My point of view has backup from two of history's greatest thinkers on the mind and the self. Way back in the first century AD, the Greek philosopher Epictetus advised his student at the time to 'First say to yourself what you would be; and then do what you have to do.'[1] Much later, William James, the father of modern psychology, said something similar, with seeming exasperation: 'For God's sake, choose a self and stand by it!' Which sounds like a challenge, to me.

So, step one: decide what to work on / choose a self. Based on things that annoy me about my own cognitive ability — and a very unscientific poll of my friends and family to see what they would improve on if they could — I have picked the following:

1. Attention — be able to stay on task and resist distractions
2. Worrying — find a way to turn down the stress
3. Creativity — learn to order new ideas on demand
4. Navigation — add a much-needed sense of direction
5. Time perception — find ways to enjoy every moment, and kill boredom
6. Number-sense — try to get a 'head for numbers' and a handle on logic

All of these are skills that I have to one extent or another but are never completely under my own control. Perhaps if I can bolster the brain regions and circuitry behind each of them, I will have a better chance of running my mind, rather than just being dragged along by it.

Step two: do what you have to do (and stand by it).

This bit is trickier, because it is tied up in the broader question of whether such a thing is actually possible. The idea that we can somehow harness neuroplasticity to take manual control of our own brains, and steer them in the direction of our choosing, is still an open question, whatever the self-help section of the bookshop would have you believe.

One answer seems to come from all of those brain-game books, apps, and websites, which have been knocking around in more or less the same format for about a decade now. Several casual acquaintances, when I told them about my mission to improve my brain, have said: 'Have you heard about such-and-such a commercial brain-training programme? My granddad / husband / friend does it and swears by it...'

True, there are a lot of these brain games about, and most are loosely based on the kinds of tests that psychologists use to measure cognitive skills in the lab. Most feature memory puzzles, mental arithmetic games, and the like, and generally test your baseline skills, then offer a daily 'cognitive workout' followed by updates on your progress. The best-known purveyor of such games, Lumosity, claims to have daily workouts that train 'skills, such as planning, logical reasoning, selective attention, and more'. They are careful not to say much more, since being fined $2 million in January 2016 when the US Federal Trade Commission ruled that the company 'deceived consumers with unfounded claims that Lumosity games can help users perform better at work and in school, and reduce or delay cognitive impairment associated with age and other serious health conditions'.[2] The ruling included a scathing assessment of the company's methods, accusing them of preying on consumer fears without the science to back it up. But, still, their website makes a big deal of how scientific it all is, and it's hard not to get the impression that all of this science will make a real change to the workings of your brain.

The trouble is, when you hear what brain scientists have to say about them, these kinds of training programmes seem like less of a good investment of either time or money. Most neuroscientists are hugely sceptical that these generic brain-training games are doing anything to the brain; when large-scale studies of the most popular of these games started investigating what they actually do, the answer was: not much. One study, of around 11,000 people,[3] found that brain puzzles and games do nothing to make you smarter in general. They do make you a bit better at the game you have been practising, but even then the effects don't last very long.

Faced with the fact that millions of well-meaning middle-aged folk were lapping up the hype (by one estimate, Americans alone spend more than a billion dollars a year on brain-training products), a group of neuroscientists recently wrote a strongly worded open letter to warn people that brain games weren't going to stave off Alzheimer's or keep them young.[4] Meanwhile, scientist-authored blogs, with names like 'Neurobollocks' and 'Neurobonkers', appeared on the scene, primed and ready to tear apart anyone who stretched the data too far for a sale or a good story. From where I was sitting, it was starting to seem as if any scientist who genuinely extolled the virtues of brain training probably had a version of their own to sell.

The other much-touted option is a monk-like dedication to meditation — which, because I don't excel at sitting still for long, I have never gotten along with (and, besides, I have always managed to kid myself that I at least dabble in mindfulness by doing yoga and stopping to smell the flowers when I walk the dog). But, lately, even meditation has started to get a bad rap, with some psychologists claiming it has a dark side, of common yet unreported side effects. One study of people who visited a meditation retreat back in the 1990s found that a small minority of them experienced panic attacks or depression.[5] Another, more recent, study found that daily meditation practice actually raised the levels of stress hormones in volunteers' saliva.[6] Neither of these outcomes sound ideal, particularly if, like me, one of the things you were wanting to change was an overactive panic button.

## FUEL AND MAINTENANCE

With all the controversy about brain training and meditation, it's perhaps surprising that there is a kind of training that definitely does benefit the brain, and it doesn't involve any thinking at all — it involves getting off your butt and moving.

Physical exercise has been shown pretty conclusively to boost memory and cognitive skills and to improve mood. In experiments where people did some kind of mental task, then either stopped and did some exercise, or sat still and had a break, people who had exercised scored far better on cognitive measures that came afterwards than the couch potatoes did. The explanation for this seems to come down to certain chemicals called 'growth factors' that are released by the body during exercise. Growth factors seem to put the brain into a kind of super-plasticity mode, where whatever you see, do, or learn is more likely to stick.

The most important of these chemicals is a protein called brain-derived neurotrophic factor (BDNF), which has the job of keeping existing neurons healthy and encouraging the development of new ones. The more BDNF that is around, the lower the threshold at which the brain commits to making new connections, and the less effort it takes for any new information to stick. Just one session of exercise has been shown to increase BDNF levels in the blood, but regular exercise is shown to be even better because it makes the brain more sensitive to BDNF. So if you exercise regularly, each time you break a sweat you get an even bigger bang for your buck.

As well as getting the brain ready to make new connections, BDNF — helped by another growth factor called IGF-1 — also stimulates the birth of new neurons in the hippocampus. This, too, probably puts the brain in a better place to learn new things, by adding more space to the brain's memory drives. Exercise also makes it more likely that blood vessels will sprout up to service the newly busy area — by boosting another growth factor, called vascular endothelial growth factor (VEGF), which specialises in building new blood vessels. And a lack of physical activity has been shown to reduce the speed at which electrical impulses pass between neurons. All in all, exercise makes for a healthy brain that is primed and ready to learn.

How much exercise you need for all of this to happen is less clear. The US government recommends 150 minutes of moderate exercise a week; 75 minutes of vigorous activity is good, but a study that investigated this found that more is almost certainly better: one recent analysis of over 600,000 people found that the best recipe for general health was 3–5 times that (an hour-plus per day if you are doing the low-impact version), but that doing up to ten times that amount didn't seem to do any harm to the body, presumably including the brain.[7]

Put together, what these results suggest is that working on your brain as if it were somehow detached from your body isn't going to work very well. Any plan to tone up the brain has to involve exercise if it has any chance of sticking there physically. Perhaps human brains evolved that way: there's not much need to invest in learning if you sit around in a cave all day, after all, but if you go out to explore, it's worth committing what you learn to memory. Whatever the reason, moving and learning are very much connected.

On the flipside, obesity has been shown to be particularly bad for the workings of the brain. Research by Alexis Stranahan, of Augusta University in Georgia, has found that, at least in mice, obesity switches on a sequence of events that turns the microglia — cells that gobble up anything foreign or unwanted in the brain — against connections between neighbouring neurons.[8] The result is that perfectly good connections start being gobbled up for no good reason. Stranahan, among others, is investigating whether this is why diabetes and obesity are so often linked to decline in cognitive skills.

A bad diet also seems to be particularly bad for the hippocampus, a key area for processing memory. This had led some researchers to conclude that a western diet, high in fat and sugar, might make it more difficult to remember what we know about eating properly, which might make you more likely to choose a bad diet in future — and so on and so on until the brain is as unhealthy as the rest of you.[9]

Too little food is also bad for your brain, for more obvious reasons. Food is brain fuel — so if there's not enough in the tank, it makes sense that the machine won't run very well. Interestingly, there seems to be evidence that getting 'hangry' (feeling angry when you're hungry) is a real thing, caused by the brain pulling out all the stops as it desperately tries to persuade us to seek out food. First, the hunger hormone ghrelin, from the gut, activates the amygdala, which makes us feel stressed, worried — or, in my household, downright furious about everything. Then the stress hormones cortisol and adrenaline are released, which prompts us to invest what energy we can muster into solving the problem.[10]

Again, if you look at it through the handy lens of the caveperson, it makes a lot of sense: when food was too scarce to go around, only the feisty, 'hangry' ones, who would club anyone that got between them and the last portion of mammoth, would have survived. 'Hanger' is a key survival skill. (That's my excuse, and I'm sticking to it.)

As for specific brain foods to eat, the short answer is: the good stuff. A diet that is high in trans fats and sugar has been linked to inflammation in the body — part of the immune response — and this has serious effects on the brain, putting it into a low-energy, hiding-away kind of sickness mode, which has recently been linked to a risk of depression. A diet high in fruit, vegetables, and oily fish does the opposite, keeping inflammation at a healthy level.

The 'good stuff' obviously contains a huge variety of nutrients, some of which have gained an almost mystical reputation as brain-boosters. Of all of them, omega-3s and B-vitamins and flavonoids seem to be among the most popularly touted as important. As with brain training, though, most of the advice out there in the public is irritatingly vague about what they actually do to support the brain, and how come most brains function fine even without sticking to an organic macrobiotic superfood diet.

Omega-3s in particular have an unstoppable reputation for their brain-boosting skills, but what exactly they are doing in there is often conveniently skipped over. That's because, as with so much about the brain, most of what we know about what happens at the level of brain cells comes from animal studies; human studies haven't quite gotten to the bottom of how adding omega-3 to the diet affects behaviour and cognition just yet.

Nevertheless, experiments on rats have shown that adding omega-3s to their diet does increase the levels of omega-3 fats in their brain tissues. Once inside the brain, omega-3s constitute an important building block of cell membranes, the barrier between one cell and another, which are built from two layers of fat molecules.

Omega-3s are not the only fats involved in building membranes, and if your body can't get enough of them then it will use other kinds of fats, like the saturated fats most of us consume plenty (too much) of. This explains why, if you don't eat any omega-3s at all, your brain can still make cell membranes: that bit had always bugged me — if they are so crucial to membrane formation, how come so many people manage to have functioning brains with no omega-3s in their diet at all? The difference is that omega-3s and other polyunsaturated fats are longer in structure and more 'kinked' than saturated fats, which makes the membranes more flexible, and better able to transmit both electrical and chemical messages between them. Electrical signals need ion channels, or gaps, which form more readily in a flexible membrane; and chemical messages need fatty bubbles called vesicles to pass them across the membrane.

In a petri dish, omega-3s have also been shown to increase the growth of connections to other neurons and to help newly grown neurons turn into mature ones — which makes me think that eating more of them is not a bad idea for anyone hoping to change their brain. The good news is that if you don't get any omega-3s in your diet, your brain won't grind to a total halt — it just won't be firing on all cylinders, and therefore won't be in the best possible condition to learn and to change.

With this in mind, I'd quite like to know if I have decent levels of omega-3s in my blood — or more correctly, a healthy balance between them and their less healthy cousins, the omega-6s. There is some evidence that what matters isn't the exact amount you have, but the ratio between the two. Finding this out isn't easy. Despite many unscientific websites telling me that lethargy, and dry hair and skin are sure signs of an omega-3 deficiency, it all sounds a bit vague and unscientific for my liking.

Companies do exist that will test your omega-3 levels for £50 to £100, but it seems like a bit of an extravagance when you already know whether or not you are getting the recommended one or two portions of oily fish a week, and there is no good evidence to date that adding more omega-3 on top of that makes any difference. As for whether supplements are as good as the real thing, the evidence suggests they are less good than eating fish — but if you hate oily fish then they are probably the next best thing. Studies of children who are already malnourished and found to be lacking omega-3s found that supplements improved reading, spelling, and school test results. Similarly, in studies of children with behavioural problems, supplements improved both their low levels of blood omega-3 and also reduced the number of tantrums and other problem behaviours. In adults, too, low omega-3 has been linked to depression, and, in experiments, supplementation seemed to reduce the reaction to stress. This effect isn't restricted to the brain, by the way, it affects the whole body: reducing heart rate, blood pressure, and the stress-related hormones adrenaline and cortisol. This is further evidence that the brain and body are very much connected — which sounds obvious, but is often forgotten in discussions about how you might change your brain.

Incidentally, some researchers think that adding fish to the diet was what allowed our early ancestors in Africa to grow such large brains in the first place. According to estimates, they were almost certainly eating more fish than modern humans do, which is an interesting thought. Maybe they had brains that were better set up for learning than ours are?

As for other brain-boosting foods, flavonoids — plant pigment chemicals found in berries, nuts, grapes, and cocoa and tea, among other plant-based foods — have some evidence to back up their reputation as superfoods. According to a recent review of what flavonoids are actually doing in the brain, it still isn't clear exactly how they work; it may be that the flavonoids, or the smaller components that the body breaks them down into, are involved in important chemical reactions underlying brain function; or it might be that they simply improve blood circulation around the body, meaning that more of the good stuff gets to the brain.

Either way, clinical trials in humans and rats have shown that flavonoid-rich diets help memory, learning, and cognitive function in general.[11] The only bad news is that eating lots of chocolate — even the really dark, bitter stuff — isn't enough to get your fill of flavonoids, and with the brain-stunting effects of too much fat and sugar, it's probably better to go with a mix of fruits and berries, with a bit of wine and chocolate on the side. Belt and braces, I always think.

Then there are the B-vitamins: B12 in particular is thought to play a role in protecting the myelin coatings on long-range brain wires, which helps with any thinking that involves more than one brain area (most thinking). A team at Oxford University is currently researching whether B-vitamin supplements help people with mild cognitive impairment, an early form of dementia, from progressing to full Alzheimer's. It's still early days, but it looks as if it might just help.

All in all, then, it looks as if any serious attempt to get the best out of the brain is going to have to involve looking after the body through exercise, just the right amount of food, and a healthy, mixed diet. But that's not all. To complicate matters further, it seems there is another factor to consider in all of this, and it isn't technically part of the human body at all. Research at University College Cork, in Ireland, has shown that there is a huge amount of communication between the gut and the brain, and that gut microbes have a big part to play in the contents of that conversation — whether the gut is healthy, stressed, or malnourished, and how the brain, and body, reacts.[12]

A lot of the research on this has been done in mice because they can be kept in a sterile environment for their whole lives — and because they are bred from other 'germ-free' mice, they don't pick up any bacteria from their mother during birth either. This means that researchers can manipulate the microbes that the animals come into contact with and monitor the effects on their brain and behaviour.

The research showed that germ-free mice live longer, but they display abnormal social behaviour and stress responses. Through some intriguing experiments, they found that adding gut bacteria from mice with normal stress response (via faecal transplant — you can learn how to do a faecal implant yourself online, but you probably won't want to ... suffice to say, you will need a new blender afterwards) makes these stressed-out germ-free mice behave normally, while giving normal mice the gut microbes of a stress-prone mouse does the opposite.

It even works cross-species: infecting healthy mice with the gut flora of a depressed human made them act depressed, in lab tests used as a proxy for depression in mice. Depression has a distinct microbial signature, the researchers say, and if you infect healthy mice with it, they start to behave accordingly. More than that, in human trials, adding a dose of a particular strain of a common bacterium (*Bifidobacterium longum*) to healthy volunteers not only reduced their levels of stress hormones and made them feel less stressed, it also slightly improved their performance on memory and learning tests in the lab.

Over the past few years, it has emerged that there is seemingly nothing that gut bacteria can't do for the brain. They regulate the birth of new neurons in the hippocampus, change levels of certain neurotransmitters, and also have a hand in the myelination process. None of this happens directly — as far as we know, there are no bacteria actually swimming about in the brain, but they seem to do something to start the chain of events that makes these things happen. One route by which they do this is via the vagus nerve, which links the gut, along with many other organs, to the brain.

As well as giving the brain regular updates on the health of the organs, and on whether it's time to eat, the vagus nerve has been implicated in driving emotions, too — particularly anxiety and fear. This is almost certainly where the expression 'gut feeling' comes from; it seems that the gut, not the brain is the first to sound the alarm in a threatening situation. In experiments where the stomach-to-brain section of the vagus nerve was severed in rats, they were far less naturally fearful in open arenas, where usually they would have kept a close watch on their surroundings.[13]

When the researchers at University College Cork gave healthy mice a dose of *Lactobacillus* and *Bifidobacteria,* the mice showed changes in stress-related neurotransmitters in various parts of the brain, had lowered stress hormones, and were less anxious under stress. If the vagus nerve was cut, though, whatever message was passing along the nerve couldn't get through, and their brains and behaviour were unchanged. No-one knows yet what kind of information the vagus nerve is sending from the gut — and its microbes, to the brain — but whatever it is, it seems to be important information for both body and mind.

At the Society for Neuroscience conference in Chicago in October 2015, John Cryan, who leads the research in Cork, said that it raises the intriguing possibility that probiotics (living doses of friendly bacteria) or prebiotics (the right food to encourage specific bacteria) could help people with depression and autism, and, potentially, to improve brain function in general. The other option is to change diet — the specific bacteria that seem most beneficial to the brain are those that help break down dietary fibre, so making sure there is plenty of that alongside your superfoods is probably a good idea. According to Tim Spector, a professor of genetic epidemiology at King's College London and author of *The Diet Myth: the real science behind what we eat,* variety is the key to keeping your gut microbes happy — although, he suggests having more celery, garlic, Belgian beer, and chocolate would be a good place to start. Spector's book aside, there is no specific dietary advice yet on how to give your gut bacteria the best environment in which to flourish. Nonetheless, the research into gut microbiomes does provide some intriguing hints that diet affects the way our brains work in ways that we haven't considered, and that we might not be as much in control of our feelings and behaviour as we might like.

Finally, there's sleep, the lack of which is seriously bad for brain function. Even missing a couple of hours of sleep a night hammers your cognitive functioning, and can leave you at risk of long-term bad health — even dementia. Science hasn't quite worked out why sleep is important for the brain, but it is widely presumed to be crucial for maintenance, repair, and memory storage; according to Harvard's Sleep Lab, most adults need 7.5 to 9 hours of sleep to do all of this. Any less than that and memory, mood, and reaction time all suffer. Only paying back the 'sleep debt' with a good night's rest will return brain function to normal.

There are exceptions, though. A rare mutation on the DEC2 gene, allows some people to get by on four to six hours of sleep with no foggy head and, as far as researchers have yet been able to tell, no other effects on their health or lifespan. Only a few people with this mutation are known to science, and if you don't wake up after a short night's sleep feeling fantastic then you probably don't have it. Researchers are working towards one day giving other people the same benefits as the 'short sleepers', but so far there's no better advice than to go to bed when you're tired and get up at the same time every day.

What is most important to take away from all of this, though, is that while it's tempting to think of the brain as the driver of everything we do, think, and feel, there is a huge amount of evidence that the body has a big hand in it, too. This might actually be helpful to know when making any plan to keep the mind in tip-top condition. From what I've seen, it looks as if running the engines of the mind has a lot to do with running the body a little better. Whether there are any specific brain exercises that are needed on top of that ... I guess we'll see.

## NEUROPLASTICITY

With so much uncertainty about pretty much everything so far, it seemed like a good idea to delve into the details of this thing called neuroplasticity to find out what it's all about.

The good news is that whatever anyone thinks about the type of brain-training games and self-help books that invoke neuroplasticity, they didn't just make it up. One thing that everyone can agree on is that if neuroplasticity isn't happening inside your brain right now, you're probably dead. It becomes problematic, however, when you try and pin down exactly what people mean when they say that something 'promotes neuroplasticity' or 'rewires the brain'.

What we do know is that if you could look at your brain with a microscope right now (and we know this because scientists have done it in mice) you would see tiny bumps growing out of the branches of your neurons, feeling around like the tentacles of a curious octopus, sometimes connecting to a neighbouring cell, and sometimes shrinking back again.

This process of growth and retraction is happening all the time, and while it might seem a bit wasteful to keep growing and shrinking all the time without much to show for it, it does allow the brain to stay primed and ready to make new connections when it needs to. Most of the time, at least in adults, there isn't a great deal to see from one day to the next — the brain ticks along, making a few connections here, breaking a few there, with not much in the way of real, wholesale change. Only when something memorable happens, or you make an effort to learn something, do the new connections start to outnumber the old and the brain begins to change.

This, of course, is the basic story that gets trotted out whenever somebody wants to convince you that you can change your brain. If they really want to hammer the point home they'll probably quote Canadian neuroscientist Donald Hebb who (almost) said: 'neurons that fire together wire together', back in 1949. (What he actually said was: 'When an axon of cell A is near enough to excite a cell B and repeatedly or persistently takes part in firing it, some growth process or metabolic change takes place in one or both cells such that A's efficiency, as one of the cells firing B, is increased.'[14] Stanford University neuroscientist Carla Shatz gets the credit for the snappier version.)

The other line of evidence for brain change comes from brain-imaging studies in human volunteers, which have shown that when a person learns a new skill, their brain changes physically, growing a bigger area to take on the new work. Eleanor Maguire's studies with London taxi drivers are probably the best known. Over the past decade, she has shown that the posterior hippocampus, a part of the brain that is involved in spatial memory, gets larger in drivers who spend more time memorising routes around London to pass 'the Knowledge' — which tests their memory of 320 routes, 250,000 streets, and 20,000 landmarks in central London.

It takes two to four years to acquire 'the Knowledge', and it's a fiendishly difficult test to pass. Maguire found that, to cope with the challenge, the brain has to invest more resources in spatial memory, sprouting more grey matter in the hippocampus. And since there is only so much space in this crowded inner part of the brain, a neighbouring area, the anterior hippocampus, had to shrink to make room. This, the same study suggested, makes taxi drivers perform worse on certain visual memory tasks.[15]

There are other examples of brains growing and shrinking in response to learning, too. Studies with musicians showed that they have larger areas of the brain associated with fine movements and sound-processing than non-musicians. These changes match up with the amount of time someone has been practising, showing that it is practice — rather than any innate *advantage* — that makes the difference. Years of practice don't seem to be strictly necessary, either. Novice jugglers show an increase in size in areas of the brain that process fast-moving objects after just a few weeks.[16]

All of these studies are well known and are done by respected scientists — who know their stuff and aren't trying to sell you anything. But when anyone tries to put what Hebb said and what these brain-scanning studies are saying together, and pretend that they know what is going on, the whole argument very quickly starts to unravel. So far, there is no way to watch firing and wiring going on at the same time as watching blobs on brain scans getting bigger, in living human brains. Which means there is no way of knowing whether the increase in brain volume seen in brain-imaging studies comes from growth of new cells, a rush of new connections, or something else, like new blood vessels sprouting up to service a busier bit of brain. All in all, it makes the 'rewire your brain' spiel a little more difficult to swallow.

This was all becoming a bit complicated, so I got in touch with Heidi Johansen-Berg, a professor of cognitive neuroscience and the head of Oxford University's functional brain-imaging centre. We'd spoken a few times over the years for various things I'd been writing. She definitely doesn't come across as prone to exaggeration, and so, to my mind, was the perfect person to cut through the hype and *tell it like it is*. I asked her to talk me through what we know for sure about brain plasticity on the phone and, despite the fact that I'd been pretty much stalking her around that time for comments on various articles, she agreed.

What Johansen-Berg told me was that it's actually pretty unlikely that new connections — the 'firing together and wiring together' bit — would account for a bigger blob showing on a brain scan: 'It sounds attractive, as something that would go on when you are learning something, but, because the size of those connection sites is absolutely tiny, it's very unlikely that an increase in spines would give you something that you could detect with MRI.'

So if new connections aren't likely to be bulking out the grey matter, what is? Johansen-Berg wanted to know the same thing, so she did a trawl through the research in this area, later publishing a review article in the journal *Nature Neuroscience*.[17] In this overview of current research, she concluded that brain change involves lots of things, but at the moment it's not possible to say which of them is behind the appearance of bigger blobs on brain scans, or (more likely) whether those growing bits of brain come down to a combination of all of them. In a nutshell, she told me that the popular idea of 'rewiring a brain' could involve any or all of the following.

### *More neurons*

In some parts of the brain, including the hippocampus, new neurons are created to cope with the demands of learning and memory. So the birth of neurons (aka neurogenesis) is probably behind at least some of the change in the London cabbies' brains. But neurogenesis hasn't been conclusively found to happen outside of a few specific areas of the brain, and so can't explain every growing blob found on every before-and-after brain scan.

### *More 'glue'*

Neurons are what most of us think of as 'brain cells'; they are the ones that carry out the fancy-pants electrical processing that (somehow) turns into our thoughts, desires, and memories. But they are definitely not the only brain cells that make up grey matter. People argue about the exact numbers, but what we do know is that neurons are at least equalled and may be outnumbered by another family of cells called glia.

*Glia* comes from the Greek word for glue — a name these cells were given because they make up a kind of sticky scaffolding system that holds the neurons in place. And for a long time that is all they were thought to do. But, recently, there have been a few tantalising clues that they might have something to do with learning as well.

One type of glial cell, called an astrocyte, in particular has attracted researchers' interest. In animal studies, where you can teach the animal something, and dissect its brain afterwards to see what changed, astrocytes have been found to become larger after an animal learns. So this might account for measurable changes in human brains, too. 'This might well be something you could see on a brain scan,' says Johansen-Berg.

It might be that, when we learn, the astrocytes make sure that a particular circuit is better serviced, and so we can get on with the job of thinking more easily. Or it could be that the astrocytes themselves are doing something more directly linked to the thinking process. So far, nobody knows. Whichever it is, astrocytes are clearly important for the job of thinking, and human astrocytes are particularly good at the job. In 2013, a group of scientists put human astrocytes into a mouse brain to see what would happen to their navigation skills. As a result, the mice became a lot better at navigating a maze and remembering where objects were hidden compared to the control mice, who only had their own astrocytes to work with.[18]

More intriguingly still, studies of Einstein's brain have shown that he had far more astrocytes in brain areas to do with abstract thinking than you would expect. So, even though astrocytes don't communicate at the lightning-speed of neurons, they might be doing something that helps us think. Or, as Johansen-Berg put it in her fabulously understated way, 'There is a growing sense that we might have been missing something quite important about the astrocytes.'

### *More pipes*

While astrocytes are busy doing whatever it is that they are doing, animal studies have also shown that the blood vessels that link them to the neurons also sprout new branches. When a region of brain is being worked particularly hard, more blood means more energy, oxygen, and all the other things that an active cell needs to keep running efficiently. As an explanation for a changing brain, new blood vessels sounds less exciting than new neurons or connections. Still, blood vessels make up around 5 per cent of grey matter, so if they start to branch out, then it might add enough bulk to show up in a scan. If that's the case, what is often sold as 'rewiring the brain' might actually be more of a re-plumbing job.

### *More cables*

Rewiring certainly does happen, though, whenever we learn something new. New spindly branches between neighbouring neurons almost certainly contribute to the growth of the brain blobs. For example, studies back in the '90s showed that people who have had more education throughout their lives have more dendritic branches, the small, localised connections between neighbouring neurons.

What most of us probably think of as 'rewiring', though, comes down to the white matter — the long-range cables connecting one region of the brain to another part, which might be several centimetres away. Almost any bit of thinking you do needs input from more than one brain region, so how well connected these different bits are, and how fast the long-range wires can conduct electricity, can make a big difference to how efficiently the brain can process information. Wires that have become connected in unhelpful ways might be behind some of our less desirable habits, like eating to excess or getting a kick out of gambling.

White matter is named for the white fatty covering, called myelin, which coats the axons of neurons, insulating them and allowing electrical signals to pass along the axon ten times faster. More electricity passing along the wires as we repeat thoughts and behaviours is what gives the brain the impetus to upgrade a normal connection to a super-fast one. For those interested in the details, it seems to work like this: electrical activity stimulates the release of a chemical called glutamate, which attracts glial cells called oligodendrocytes. These cells get to work, adding a spiral of fatty insulation, made from the neuron's cell membrane. More activity in a circuit can also lead to changes in the wiring by making the wires themselves longer, fatter, or more densely packed.

Once a coat of myelin has been added to the axons of the neurons, the extra layer of insulation inhibits branching, protecting well-used highways from being diverted or broken down. This is another reason why it is so difficult to unlearn bad habits.

This particular mechanism could cause a bit of a problem with any attempt to rewire my brain. If the pathways I want to change have been there long enough to become properly entrenched, with a nice thick coating of myelin, will it still be possible to change them? To make things worse, these wires apparently don't just snake randomly around the brain, making and breaking at a moment's notice — they are bundled up into thick bunches of cables, called fasciculi, which keep them all neatly together and running in the right direction (see figure 2). Imagine the effort it would take to unravel that lot and then make a few tweaks! It just doesn't seem feasible.

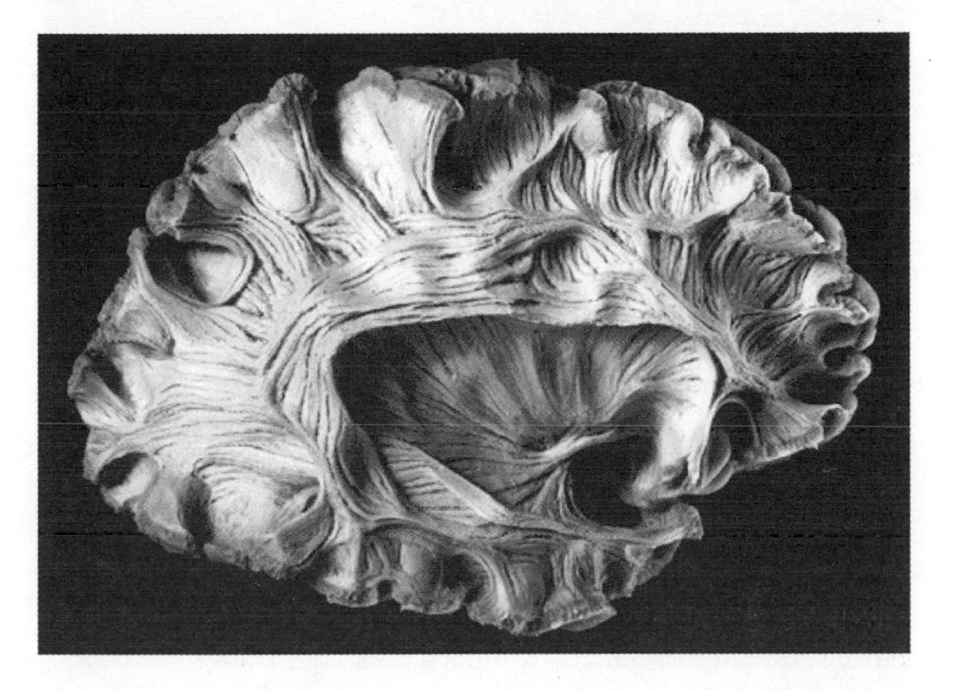

Fig 2. The fasciculi of the brain. (Courtesy of Lippincott Williams & Wilkins)

Heidi Johansen-Berg tells me, though, that adding new branches to the existing long-range cables running between brain regions might be possible: 'There is much less evidence for that but there is some evidence.' In one study, from 2006, where macaques were taught to use a rake to bring food towards them, they sprouted new connections from visual areas of the brain to the areas involved with knowing where their limbs are in space.[19] The scientists who conducted the study don't claim that a new wire was added to the bundle, though. Instead, they say that a new branch probably shot out from a wire somewhere nearby.

A less drastic option for the brain is to tweak what is already there — not so much changing the wiring as the way in which the wires are used. Neuroscientists distinguish between structural changes — actual physical changes to the system itself — and functional changes, differences in the way that that structure is used, electrically or chemically, or in the strength of the connection at the synapse, where two neurons meet. Both structural and functional changes can make a big difference to the way the brain works in the real world, and one kind of change may lead to the other.

All in all, it's fair to say that neuroplasticity is fascinating but complicated, and even the experts don't know exactly what is going on inside our heads when we learn something new. There are good grounds to be confident that neuroplasticity is a real thing, and hasn't been made up by unscrupulous marketing people. But, on the other hand, if someone tries to tell you that doing any one thing will make your neurons fire together, wire together, and, *hey presto*, will rewire your brain, they are being about as honest as those web ads that promise to erase your belly fat if you follow their 'one weird trick'.

I must admit, when I learned this, I was a bit worried. How were changes in my brain to be measured, in the labs of the great and good, when not one of them really knows what is happening to it? Short of having bits of my skull replaced with glass and strapping a microscope to my head, there is no way of knowing for sure. On the other hand, if I can feel the changes, and the scientists can measure by other means — like before-and-after cognitive tests, or changes in electrical activity — then it's a fair bet that something physical has changed or is in the process of changing; I just won't be able to put my hand on my heart and tell you what. Which sounds like a cop-out, but it's really not; no one else could tell you either — and if they say they can then, at best, they're exaggerating.

So, this much we know: the brain's circuits are a bit like muscles, in that if you give them more to do, they will get stronger and work better. So brain training *should* work, right? This, too, is a contentious area, with scientific analyses flying back and forth between academics, and the media trumpeting that brain training does or does not work, at every turn. It's no wonder that everyone is so confused.

One thing that bothers me is why it's so difficult to tell if brain training works or not. If my mission were to tone up my 40-something body, it would be easy. If I wanted rock-hard abs, I'd take up Pilates or commit to doing sit-ups every morning before breakfast; if I managed to keep it up for a month, I'd begin to see a much less spongy belly. If I wanted to improve my general fitness, I'd go running every morning instead — and, after a month, I'd know I was improving if I felt significantly less like death while doing it.

In the brain, what with it being locked in a box, success is harder to measure. The best anyone can do to measure such changes is to point at differences on a brain scan or at before-and-after measurements on standardised tests — which sounds simple enough, until you consider that most cognitive abilities involve widely distributed networks in the brain, and there may or may not be small changes all over the place. Add this to the fact that the brain is changing all the time anyway, and it's no wonder there isn't a simple answer.

To be fair to the neuroscientists who are working in the field, the challenge is huge. Anyone who wants to prove that a particular brain game does anything useful needs to show that any changes in score are down to the training, and not due to a cunning new strategy devised between the initial and follow-up tests, or because something totally unrelated has changed in a person's brain during the same period. Anything like finding a new love, learning a new chess strategy, or having some kind of unexpected life stress might make as many, if not more, changes as a few minutes of training each day.

In children, it's even more difficult to say what drives changes, because they are growing and changing even more than the rest of us. Cognitive neuroscientist Dorothy Bishop, of University College London, who has a healthy scepticism about brain training in general, points out that, if you measure children's feet before and after a course of brain training, they will probably have grown. Does that mean that brain training makes children's feet grow? Probably not.

Then there is the not-insignificant 'Hawthorne effect' — a kind of variation of the placebo effect, which says that if you give people a lot of attention and they know that they're being observed, they will get better at whatever it is that you are observing. So it could be the warm, fuzzy glow of being the centre of attention that causes an improvement rather than anything magic about the task itself. (Like if you spend a year thinking almost exclusively about your own brain, for example. *Hmmm.*)

Finally, there is the question of transfer: does the training make you better at just this specific game or does it help with other things, too? This is the real nub of the issue, because while it's fun to get good at a brain game on the computer, most people aren't doing it for just that reason: they want a better memory or reaction time or reasoning skills *in real life*. And whether transfer is occurring is not easy to prove.

Going back to the body-improvement analogy, transfer is a simple enough thing to spot following physical exercise. You wouldn't do sit-ups and expect to get toned biceps or a pert butt, because it's obvious that sit-ups mainly work the abs. The benefits don't transfer to any other bit, especially not the bit you are lying on at the time. You might expect, though, to get toned arms *and* stronger abs if you did 100 press-ups instead, because you can't do press-ups without keeping your body straight, and that works the abs as well as the arms.

Something like running covers even more bases — because that works not only several major muscle groups, but also the heart and lungs, so not only will it make climbing the stairs easier, but swimming and playing football with the kids will get easier too. Likewise, swimming or playing football will make it easier to run. Cardiovascular fitness training transfers further than working one or two areas of the body; it improves the underlying basis of health.

Studies like Maguire's (on taxi drivers taking 'the Knowledge', mentioned earlier) have shown that by taxing particular parts of the brain, you can tone up specific parts of the circuitry. As long as you keep up the exercise, the circuits will become stronger, work better, and you will be able to do more with them. But at no point has Maguire found that taxi drivers get cleverer in general. The changes to the taxi drivers' brains are the equivalent of having rock-hard abs after doing a lot of sit-ups (not something many taxi drivers can lay claim to). What brain-training programmes aim for is the brain equivalent of cardio training — some way to work key, underlying brain skills that will make everything work better.

This cardio-fitness equivalent for the brain is proving difficult to come by. One of the best-studied candidates is working memory: the ability to hold information in mind while you manipulate it to work out what to think, say, or do next. It's a skill that has a hand in pretty much every complex task that the brain has to do, which is probably why differences in scores on working-memory tests predict general intelligence and reasoning ability. If it were possible to improve how working memory functions — the thinking goes — it might just make the whole system work better, and the person who owns that brain a little bit smarter.

It does make sense, which is why there has been a huge amount of research into whether working memory can be trained and, if so, whether that makes people smarter in general. The trouble is, while some studies have found that it works a treat — improving other measures of intelligence and brain function — others, even exact reruns of previous studies, have concluded that it does nothing at all. All of which has led some psychologists, like Charles Hulme, of University College London, to suggest that it's all a bit of a waste of time.

Hulme and his colleague Monica Melby-Lervåg, of the University of Oslo, in Norway, recently analysed the results of over 80 studies of working-memory training, many of which claimed to have shown improvements in general intelligence after just a few weeks of training. They concluded that, when you put all of these studies together and analyse them as if they were one big study (this is called a meta-analysis), the effects on general intelligence disappear. All you are left with is a small, and very temporary, improvement on playing the game used for training.[20] Which would suggest that working-memory training is not the cardio-fitness-for-the-brain kind of exercise that we've been hoping for, or indeed been sold by brain-training companies.

'The bottom line is that all of this stuff is complete twaddle and people should forget about it,' Hulme told me when we chatted on the phone. Thankfully, he wasn't talking about the whole idea of changing your brain — that, he agrees, does seem to be possible, so long as you choose the brain skill to be improved carefully. 'Train a task and you'll improve on it — but that's not what these people are claiming,' he says. 'They're saying "train one task and everything else gets better". And that's a) deeply implausible, and b) based on the evidence, simply not true.'

Susanne Jaeggi, of the University of California in Irvine, disagrees. She has done her own meta-analysis — and over Skype, she told me that she has found quite the opposite, with a small but significant improvement on lab measures of general intelligence after working-memory training.[21]

'It looks more and more like there is indeed transfer to other skills that rely on working memory, such as reasoning, reading comprehension, even math problem-solving tasks,' she said. In fact, anything that needs working memory seems to benefit. 'But whether that is seen in school grades, that's another question. We have very limited data on whether improvements on this training translate into real-life performance.' For anyone planning to dedicate 20 minutes every day to doing these kinds of games, this is what we really need to know. It almost doesn't matter to us that scientists like Hulme and Jaeggi are arguing over whether the tiny improvement is real or not. Quite honestly, if it is so difficult to tell, under very controlled laboratory conditions, whether anything has changed, would anyone have the slightest hope of noticing any changes in real life?

As an aside, Joe DeGutis, a cognitive psychologist at Harvard University who spends a lot of his time trying to make cognitive-training programmes that actually work, told me that there is a standardised test that scientists can use to see whether real-world changes have occurred after intensive cognitive training.

A few years ago, he told me, he was asked by a funding body to translate his results with stroke patients into a measure that made sense in real life.

'They said, "I think you should time how long it takes the stroke patients to make toast,"' he tells me, looking incredulous about the whole thing. The idea does sound a bit daft, I admit — and, anyway, won't the toast just pop up when it's ready? It sounds like it might tell you more about different makes of toaster than brain change.

'Ah, well,' he said, 'there's a certain toaster that everyone uses for the toaster task; then you time how long it takes to butter the toast, and so on ...' He laughs, because it does sound a bit ridiculous. But, he said, the whole point of this kind of research is to show that it translates into real-world changes. Becoming better at an on-screen game alone isn't going to change anyone's life, no matter how much fun it is.

Even if cognitive training does improve your toast-making skills, there is another complication: the same training might not work for everybody. In her analysis of working-memory training studies, Jaeggi found that people who started off with the lowest skills improved the most. People who were measured as having a 'growth mindset' — in other words, who believed that it was possible to improve on their current skills — got more benefit from the training than people who thought that they were either inherently good or bad at something. Interestingly, people with a growth mindset saw benefits from the placebo training, too — which wasn't supposed to change anything. This suggests that, to some extent, just thinking that you are doing something to change your brain is enough to make it so.

People who sign up for training studies out of the goodness of their hearts also seem to get more benefit than people who are there for payment or course credit. Also, a laboratory measure of how much you like doing difficult puzzles ('need for cognition' in psychologist-speak) also predicted how much benefit people got from training.

As for the scale of the improvement, Jaeggi's latest analysis estimates that working-memory training might improve general intelligence by 'about the equivalent of 3–4 points on a standardised IQ test'.[22] Which, Charles Hulme told me, would be quite a significant improvement, if only it was a reliable result that always showed up when people did this kind of training. For her part, Jaeggi said that it is indeed a real improvement. 'Doubters seem to be hard to convince. There are some people who don't seem to be convinced, but I'm okay with that,' she said with a shrug.

Which looks a lot like stalemate to me. Joe DeGutis shares my infuriation. 'If the cognitive-training people worked on one problem for a while, and others could work on the generalisation process and how you take this into your life, and another group would work on another task ... At the moment, they're smooshing them all together and no one's getting anywhere,' he says. 'Half the studies are trying to get the effects and half are showing that the effects aren't real.'

Whether this happens or not, the main take-home message for me so far is that any changes occurring in the brain are so tiny that you need very sensitive lab tests — or a certain kind of toaster — to spot that anything has changed. Which doesn't sound like the kind of change that is worth 20 minutes of your time every day. And with changes that small, it is even trickier to say what may or may not have changed in the brain. For what it's worth, Charles Hulme reckons it doesn't matter anyway. 'You don't need to know what's happening in the brain — training doesn't affect the behaviours that these people want to affect,' he said, sounding slightly exasperated by the question.

Either way, for now it seems that anyone with ambitions of a new, improved brain would be better off picking one skill at a time and focusing on that, instead. Mental jogging doesn't exist yet, and chances are that it might never exist. Which brings me to the first conclusion in my mission to change my brain: *pick what you want to change.*

So it's just as well that I'm not out to improve my general intelligence or even develop superhuman powers of recall. Improvements in real-world skills that will directly affect my life sound a lot more doable. And, happily, each of the areas I have chosen to improve has a community of neuroscientists and psychologists working on ways to understand what makes these skills work and how we might enhance performance. By choosing these select skills over a more general 'get smarter' ideal, I have at least a fighting chance of getting some improvement on each particular area. How specific any improvements turn out to be remains to be seen, though. It isn't clear to me whether a London cabbie would be better than average at navigating around New York, or whether his or her superior spatial knowledge is specific to London. And if I can learn to control my fears about, say, a sudden disaster befalling my family, will all my other neuroses melt away too?

The answers to these questions — and, more importantly, whether it's worth the hassle to try to change your brain and keep it changed — remain to be seen. But, as one neuroscientist commented when I listed all the aspects of my brain that I hoped to change: 'Wow. At the end of this, you'll either be Superwoman or completely messed up ...'

Either way, it should be interesting, right?

# PART ONE

# MEETINGS WITH MY EXECUTIVE CONTROLLER

CHAPTER 1

# The Taming of the Butterfly

*'Concentration is the root of all the higher abilities.'*

— BRUCE LEE

There's a lot about the brain that nobody knows for sure, but one of the few things that we do know is that focusing attention is about the most important thing that it does. Attention is the filter the brain uses to decide what is important and what can safely be ignored — and, without it, the lessons in front of your eyes won't get translated into real physical change, at the level of neurons and their connections to each other. So, if I want to make any meaningful change to my brain, good focused attention is going to be absolutely crucial.

Which might be a problem, because leaving my luggage unattended in airports is just one example of the space-cadet tendencies that earned me the nickname 'butterfly brain' when I was about eight years old. And this kind of thing still happens all the time. Recently, I was walking along the high street in my town when I looked over and saw my friend doubled up with laughter on the other side of the road. 'You look like a crazy lady, staring up at the sky and wandering aimlessly!' she said. Charming. But she's right: when I zone out, I *really* zone out.

When it comes to getting my work done, this zoning out really doesn't help. I work at home, alone but for a very distracting dog, and — in theory — in short, intense bursts while my young son is at school. Occasionally, this arrangement works perfectly, and I end the day feeling like Superwoman. More often than not, though, I spend the day flitting from one thing to the next, doing nothing of any use at all — and although all I have to do is read a few scientific papers and send a few emails, I don't manage any of it. A few hours later, I am stressed, frustrated, and have even more to do the next day.

In brain terms, a lack of focus and a susceptibility to procrastination are two sides of the same coin — they are both hallmarks of a brain that is not under proper control of its owner. And I'm not the only one who struggles with this problem. In one recent survey, 80 per cent of students and 25 per cent of adults admitted to being chronic procrastinators. Although we like to kid ourselves that all of this makes us more creative, the evidence suggests that it actually leads to stress, illness, and relationship problems.

Apart from anything else, letting the mind wander off doesn't seem to make us any happier. In another study, researchers interrupted people during the day to ask what they were doing and to score their level of happiness. They found that when people were daydreaming about something pleasant, it only made them about as happy as they were when they were on task. The rest of the time, mind-wandering actually made them less happy than they had been getting on with their work.[1]

It was while I was banging my head on the desk in frustration one day that I remembered Joe DeGutis, a neuroscientist at Harvard University. We had spoken a few years before, for an article I was writing, and I knew that cognitive training in general, and attention in particular, was very much his thing. So I emailed him to see if he might be able to help sort me out. As it turned out, he and Mike Esterman, of Boston University, had been working on a combination of computer-based training and magnetic brain stimulation (also known as TMS) to help people focus better. So far, they had found that their programme seemed to improve people's ability to sustain attention. Like in most neuroscience studies, they had only tried it on people with serious problems — from brain injuries, strokes, post-traumatic stress disorder (PTSD), and attention deficit hyperactivity disorder (ADHD). I couldn't help but wonder: might it work for me too?

Probably not, came the reply. It's one thing to improve a brain from clinically hopeless to about average, but getting from a bit below par to a bit over is not an easy thing to do — or measure. But Joe and Mike humoured me anyway, and sent me a link to a pared-down online version of the sustained-attention test they use in the lab. They also sent over several questionnaires, which would measure things like how often I make silly mistakes because I'm not paying attention (quite often); and a 'mindlessness' scale, which would measure how much I wander around in a daze (a lot).

I sent it all back to them, and the next day the cold, hard truth hit my inbox. I had scored 51 per cent on the attention test — a good 20 per cent below average. And the questionnaires were pretty telling too. 'Considering all your results, it's very clear that you have issues with attention and distractibility both in the lab and in daily life,' wrote Joe. Then, to soften the blow, they invited me over to see if they could help. No promises, but they'd do their best.

A month or so later, I arrived at the VA Medical Center hospital in Boston, where Joe and Mike have been running the Boston Attention and Learning Lab since the year 2000. The VA is part of the US government's Department of Veteran Affairs, a body that provides lifetime healthcare for American soldiers. Many returning war veterans have difficulties in sustaining attention, so there is a steady stream of willing volunteers for Joe and Mike's studies. Post-traumatic stress is particularly problematic. When people spend their lives in a heightened state of anxiety, their attention is scattered all over the place, constantly looking for danger, leaving nothing in the tank to focus on any one particular thing. Head injuries can cause similar problems, as can strokes; because attention uses so many different brain areas, if something goes wrong with the brain, the chances are that attention will suffer. And since so many other brain skills — including memory, reasoning, and even holding a train of thought while you do a task — are built on the foundations of controlled attention, losing it can be seriously debilitating.

Walking into the VA hospital is a unique experience, to say the least. I'm not sure where the UK keeps all its war veterans, but a lot of the American ones seem to be here, under a portrait of Barack Obama, and with the American flag flying overhead. There is a real mix of ages, from younger people in wheelchairs who look like they sustained their injuries in recent conflicts, to those who look about the right age to have fought in Vietnam. Many of them are proudly wearing veteran caps and T-shirts, and sit in the lobby comparing experiences as they wait for their appointments. There must be more stories in these four walls than anywhere else in the city — and I'm dying to hear a few, but it doesn't seem right for a cheerful Brit to butt in and start asking questions about things I know nothing about.

So I wait with them and try not to look too out of place. Luckily, it's not too long before Mike and Joe appear and show me to their office upstairs. 'It's quite an interesting place to work,' Mike says, as a chatty veteran in his 80s, who isn't making a lot of sense, wishes us a good day and gets out of the lift.

What they call the 'Mike and Joe show' is soon in full swing, and I can tell that it's going to be a fun week. Joe is energetic and positive, talks fast, and bounces around like someone who has way too much mental energy on board. He later tells me he needs his own research as much as anyone. 'It's not like we have all the answers,' he says. 'We're kind of doing "me-search".' Mike is more reserved, but just as enthusiastic, and he's diligent about making sure that I have signed all the correct disclaimers before we get going, and occasionally reining in his exuberant partner-in-crime. In this particular double-act, Mike has reluctantly landed the role of 'the sensible one'.

Get him on the subject of brain stimulation, though, and he really perks up. They take me down to the room where we'll be doing the brain zapping — a disused hospital room that was last decorated in bright orange around 50 years ago. There is a huge black chair where the bed should be, an ancient X-ray viewer, and two clocks that clearly haven't ticked in years.

The chair is part of the TMS machine, which they will be using to zap me in the head the following day. And Mike can't wait to give me a demonstration of what it can do. 'It's fascinating,' he says as he moves the magnet over the motor cortex of his own brain, which controls movement, and watches his hand twitch involuntarily. 'This is how we test that the machine is working. Sometimes I come down here and do it just for fun,' he adds with a grin.

Watching the machine take control of his brain and body, I realise that this is the perfect demonstration that our every move and decision comes down to pulses of electrical activity in the brain. Of course, we all know this, but there is something unsettling about seeing that system being hijacked in front of your eyes.

Soon, it'll be my turn. But first I've got two hours of assessments to get a baseline measure of my skills — or lack of them — in this particular week, and a brain scan so they can map out which region they want to stimulate.

First comes the full version of the sustained-attention test I did at home — a test that Mike has affectionately nicknamed 'Don't Touch Betty'. My task is to look for the only female face ('Betty's') among a constant stream of male faces as they fade into each other at the rate of about one a second. When I see a male face I press a button, but when Betty pops up I 'don't touch'. It sounds easy, but the faces are all black and white and are surrounded by black-and-white scenes of mountain- and city-scapes, which fade into each other at different speeds to the changing faces.

The test lasts for 12 minutes, but it feels much, much longer. I'm finding it not so much difficult as physically impossible. Even when I spot Betty, there never seems to be enough time to tell my hand not to press the button. I spend the whole 12 minutes berating myself, as Betty's Mona Lisa–smile starts to look more and more mocking. In fact, I'm convinced that Mike and Joe will wonder if I even understood the instructions when they look at my results. I quickly assure them that I did before we move on to the next tests.

There are several tests, each of which measures a different aspect of attention. In one, called 'blink', a stream of letters flashes by at top speed. I am supposed to spot the two numbers that are thrown in among them. This is a test of brain efficiency — of how quickly my attention circuitry can re-set itself and spot something new. My guess is *not very,* judging by how often my guess at the second number is a total stab in the dark.

Other tests, though, seem less difficult. In one, I have to click only on the pictures of whole apples, and ignore pictures of apples with a bite taken out of them. Then I have to click on a dot as it jumps over the screen in an iPad version of whack-a-mole. Just as I'm thinking 'this is too easy', Joe says something mysterious about working out which side of my brain I tend to use more to pay attention. I can't help wondering if the reason why it seemed easy was because I'd missed something really important. They're tricky, these cognitive psychologists.

Then comes the final test: this time, of my visual attention skills. This test is designed to measure how easily I am distracted by something in my peripheral vision, like an email notification on the screen or a bird flying past the window. I think I do okay on this one, but by now I am starting to flag, and am sitting at the desk with my head in my hands. It's late in the afternoon and lying down for half an hour for a brain scan — even in a noisy MRI (magnetic resonance imaging) scanner — seems like a lovely idea.

We head downstairs to a white room where the MRI scanner lives. The receptionist passes me a form that probably only exists in military hospitals; I sign to confirm that there is no shrapnel in my body or metal in my eyes. 'Oh, and you'll have to change into paper pants for the scan,' Mike adds, by way of an aside. 'I hope you don't mind?' I finally click that he's not talking about my underwear when, ten minutes later, the technician hands me an enormous paper trouser suit. Phew. With that, they pack me into the scanner with earplugs and a foam neck brace, and I promptly start to doze.

The point of the scan is not to track my brain's activity but to get a 3D picture of my brain; this is then used to pinpoint the areas involved in focusing attention that they want to target with TMS. Their main area of interest is called the dorsal attention network, which links thinking areas towards the front of the brain to the parietal cortex, which lies above and slightly behind the ears and acts as a kind of switchboard for the senses.

Both sides of the brain have a version of this network, but real-time imaging studies suggest that most of the work is done by the right side. People who struggle to sustain attention, though, often have more activity in the less efficient left side.

Mike and Joe later tell me that part of their plan is to use the TMS machine to temporarily turn down activity in the left side of the attention network, which will force me to start using the right. It's sort of a brain equivalent of strapping down my dominant arm to force me to strengthen the weaker one. And since the brain will always route messages through the easiest path, once that more efficient system is working properly, it will hopefully be there whenever I need it.

The other part of the plan is the training itself — Joe's area of expertise. I am going to do three lots of 12-minute training, twice every day for a week, with extra sessions on two of the days following the brain stimulation. I start that evening, and Joe promises to email me three sessions that I can do when I get back to the hotel.

To be honest, while potentially useful, the training is pretty boring. Much like the Betty test, there is a target image on the screen that you don't press the button for — say a white cup on a brown table — but that you do press for any other cup–table combination. On my first attempt, I get only 11 per cent of the 'don't touches' correct. I don't have anything to compare it with, but it seems a fairly terrible score. Later, I'm told that, for the training to work, they need to adjust it to a point where I get 50 per cent correct. Only then can they start to improve my skills. There's certainly a lot of work to do, and there are now only four days to turn things around before I fly home to the UK. But right now, all I can do is sleep.

Next morning it's zapping day, and I wake excited about seeing my brain for the first time. I'm more than a little nervous about the actual process, but there are some benefits to being easily distracted: I forget all about my nerves when I arrive at the VA hospital and find a singalong in full swing in the lobby. A youngish guy with long curly hair is strumming his acoustic guitar, asking the assembled veterans for song requests. Suddenly, an old veteran, dressed from head to toe in red, white, and blue, jumps up and requests 'Stand by Me'. Then he bursts into song at the top of his lungs, while the guy on the guitar tries to keep up. At one point, the older man even adds a solo played on the comb-and-paper. I'm disappointed when Joe and Mike appear, looking amused to find me singing along, and drag me away to begin work on my brain.

We head up to the room with the big chair in it — and there it is, large as life, on an enormous screen. My brain. I'm not sure what I was expecting, but it all looks fairly normal; lumps in all the right places and no obvious holes that shouldn't be there. Mike has overlaid the position of the dorsal attention network with the image of my brain, and has marked a target over an area of the network called the 'frontal eye field' — which is where he is going to target the stimulation.

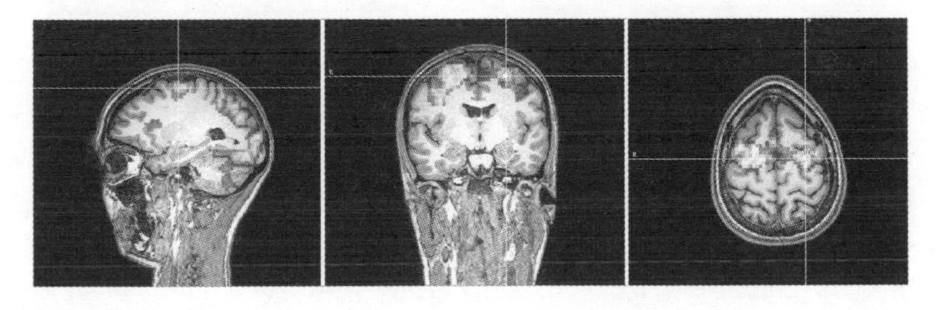

Fig 3. X marks the spot: my brain, with the brain stimulation target in the crosshairs.

I had assumed that the plan would be to jolt my brain into working better, but it turns out to be the opposite. The plan is actually to knock out activity in part of the left side of the attention network, using targeted magnetic pulses, to force me to use the right — which is the one I should be using anyway. The effect of the pulse wears off after a few minutes, so I'll do my first bout of training straight after being zapped. Then we'll repeat the process another two times. The idea is that the right side should eventually become strong enough to take over from the left in everyday life, in a self-perpetuating cycle.

The time comes to be strapped into that chair. I look ridiculous, wearing a tight headband with what looks like a coat hook on the top. This, apparently, is a bit of kit that allows them to link the brain image on their screen to my actual brain in real life, to ensure that they zap the right spot. Their research assistant, Hide Okabi, tells me it's basically an expensive version of a Nintendo Wii, linking your movements in real life to what is happening on the screen version of you. So when Mike moves the magnet over my head, he will be able to see on the screen where in my brain he is aiming.

First, he does the finger-twitching trick on me. It takes him a while to find the right spot on the top of my head — apparently everyone is slightly different — but eventually he gets it. It feels seriously weird; my hand clenches in front of my eyes as if someone is pulling it with strings. It doesn't hurt, but the hit of electromagnetic energy feels a lot like someone flicking me pretty hard on the head once a second, in time with a loud click. At first, it's just a light tap, but five minutes later and it has become very, very annoying, and I'm starting to feel a bit lopsided. I'm supposed to have three 8-minute sessions of stimulation, but when I admit to feeling a bit dizzy, they decide that two is probably enough.

After the zapping, I do another session of Joe's training and it soon transpires that, with or without brain stimulation, the training is just as frustrating after a good night's sleep. I can see my hand moving towards the spacebar in slow motion but am still unable to stop it.

After my first bout of stimulation, I do even worse at the tests, and I can tell that Mike is a bit perturbed. He's not saying much, but it seems that he expected me to do better after a short, sharp zap. But then Joe says that it could mean that I usually lean on the left side so heavily that without it I can't do the task at all. If that's the case, then I should start to improve with training.

Except that doesn't happen. By day three, I am getting no better, and I am so frustrated I feel like yelling every time I hit the spacebar in error. I feel so stupid — I have no trouble spotting the target, in fact, I see it straight away. But it feels as though a gun to my head couldn't stop me from pressing the space bar. Mike and Joe are looking even more frustrated than I am — and a bit worried too. Joe later confesses that he was scared I'd go home and write about how his training programme is 'the dumbest thing ever'.

But then, out of the blue, on day three, sometime between morning and evening training, something clicks. My score jumps from between 11 and 30 per cent correct 'don't touches' to 50–70 per cent. What's more amazing is that I'm starting to enjoy it. And when I accidentally make a mistake, I have a strange understanding of where my mind went. I realise, for example, that I got one wrong because I was wondering what my son was up to back home. And should I have wine after training, or beer? By the next day, I am doing it one-handed, while casually grabbing for my cup of tea with the other hand. Crucially, where before there had been white noise and no way of getting my mind to behave, I now feel only a Zen-like calm with just a few ripples of distraction.

Joe seems excited by this — and tells me it could be an important development. Being aware of what you are thinking is known in psychology as 'meta-awareness', and it's essential if you are trying to spot mind-wandering before it takes you too far away. 'Everyone that the training has helped finds that they get to the stage where they are a little more meta-aware. They are doing the task and they see themselves thinking about other things,' he says.

And it might be my imagination, but I feel a lot calmer, too. Normally I procrastinate when I should be working, and, as a result, I end up working — and stressing — in what is supposed to be relaxation time. But this week I'm a journalistic ninja: I do all of my work in the allotted time slot, then have a great time catching up with friends in Boston, enjoying being back in one of my favourite cities, with no guilt or stress about work not being done. It might not sound like much, but it's a revelation to me. Maybe life doesn't have to be so stressful after all.

I've also noticed a few more subtle changes outside work. On day two in Boston, I moved out of the hotel and went to stay with friends for the rest of the week. This would usually have caused me to get in a right old state — even with good friends and close family, I am an anxious house guest. I can't relax because of the nagging feeling that I'm in the way, that I should be doing more to help with the cooking and cleaning, or that I should be making better conversation — and then I'll worry that my worrying is stressing everyone out. But not this week. This week it's *all good.*

I haven't mentioned this to Mike and Joe — I have the feeling they think I'm a bit unusual — but I'm intrigued to know what has changed. But aside from confirming that there is room for improvement, and that I probably don't have ADHD, they're saying nothing until I resit the 'Betty' test on Friday. They're not just being mysterious. Keeping experimental subjects in the dark is part of how science works: the more I know, the more likely my expectations will skew the results. All I know is that training has now become easy and that the butterfly feels under control for the first time in a long, long time.

Almost 30 years, in fact. Now I'm thinking about it, I came up with a few strategies to get the 'butterfly brain' through school. I'd forgotten about these strategies until I recently found that psychologists and neuroscientists were suggesting similar things, although this time based on proper studies of the attention span. The fact that attention span is limited might seem obvious enough, but psychologists have spent decades arguing in long-winded journal articles about precisely why. When we were about 11, my friend Anna and I came up with a system that helped us get through maths class. We would do 20 minutes of work, then break for a five-minute chat. It worked a treat, and we would always manage to stick to the schedule well enough to finish our work before the end of the lesson.

Regular breaks are often touted as a way to help focus. But why should they work? Some put it all down to regular 'alertness cycles' in the brain, which supposedly give us about 90 minutes of alertness before zoning out for a while to reset. This idea has been knocking around since the 1970s — and since then, studies of productive people have backed it up, showing that they often naturally choose to work intensely for 90 minutes, three times during the morning, with a 15-minute break in between. It sounds good, but when I tried it, I found that it was not the easiest thing to fit into a working day. It was also hard to work out when my most alert time was; on some days, the alert part of the cycle stubbornly refused to show up. However, the basic take-home message is that we need to take breaks — preferably five or six a day, according to the 90-minute rule.

Other psychologists argue that it is more likely that our limited span comes down to simply running out of steam. By this reckoning, we pay attention quite literally using a reserve of mental energy, eventually running out of funds. In this view, the break doesn't have to be 15 minutes — in fact, one study found that a distraction lasting a couple of seconds did the trick. The catch is that it has to be a total distraction, like trying to do a difficult bit of mental arithmetic. Chatting to your neighbours in the office probably won't work unless their conversation is truly engaging. I can't remember what Anna's conversation was like at school, but I'm fairly sure it involved boys.

Taking breaks isn't the only strategy I stumbled upon at school that has since been backed up by proper science. Through school and university, whenever I needed to remember facts, I'd write and rewrite my notes, over and over, using different colours on each line. Or I'd change my style of handwriting for each section, or even do some bits in mirror writing. I didn't think much about it at the time, but what I was doing was finding ways to hold my attention for long enough to get the stuff to stick.

Then, a few years ago, Nilli Lavie, a cognitive neuroscientist at University College London, came up with the 'load theory' of attention, which suggests that we should deal with our limited attention not by reducing what the brain is processing at any one time, but by giving it more to do. This has since been borne out by experiments that measured how much an on-screen distraction slows people's performance while they are doing a mental puzzle. Lavie and her team have shown that the busier the screen is during the task, the easier people find it to ignore the distraction. It sounds counter-intuitive, but the idea is that if our senses and perception are full of things that we should be thinking about, we will have no room for any mind-wandering. It seems to work with all senses, too, Lavie says; so making what you are doing more mentally demanding, either by deliberately working in a noisier environment or by colouring in revision notes like I did, might be worth a try.

The only problem with making this work in the real world, however, is ensuring that all the slots are taken up with something useful and not with distractions that take your attention even further away. If colouring your notes becomes too elaborate, as I have often found it does, they can easily turn into a work of art rather than an *aide-mémoire*. And mirror writing tends to make note-taking and reading slower, which probably defeats the object. Even achieving the right level of background noise while working in a public place is never easy.

The final explanation for why attention is limited, though, is the one that makes the most sense to the adult me. Mindlessness theory says that loss of attention happens when the brain gets so used to the task that it shifts into automatic mode, essentially taking focused, effortful attention elsewhere. Allan Cheyne, a psychologist at the University of Waterloo, in Canada, is a proponent of this theory, and he suggests that some kind of alert system — maybe something as simple as an alarm every 20 minutes — could help to drag you out of dreamland and back to the task at hand.

For me, the trouble with all of these solutions is that they require not only a certain amount of mental control but a lot of organisation, and that's not always easy when you're already battling a limited attention system. And even when working at home at your own pace, these kinds of rigid schedules aren't always workable. Plus, what I'm trying to do in Boston is to make such strategies redundant by fundamentally changing the way my brain uses its attention resources. If I can do that, then I won't need to colour my notes in anymore because my brain will be much better at slipping into the zone and staying there.

So did all that training and zapping really do anything to my brain? Well, the short answer seems to be yes. My score on the Betty test showed a huge jump, going from 53 per cent incorrect before training (worse than any other healthy person they had tested, and in the region of a brain-injury patient) to 9.6 per cent afterwards (almost as good as the best healthy person in the same study).

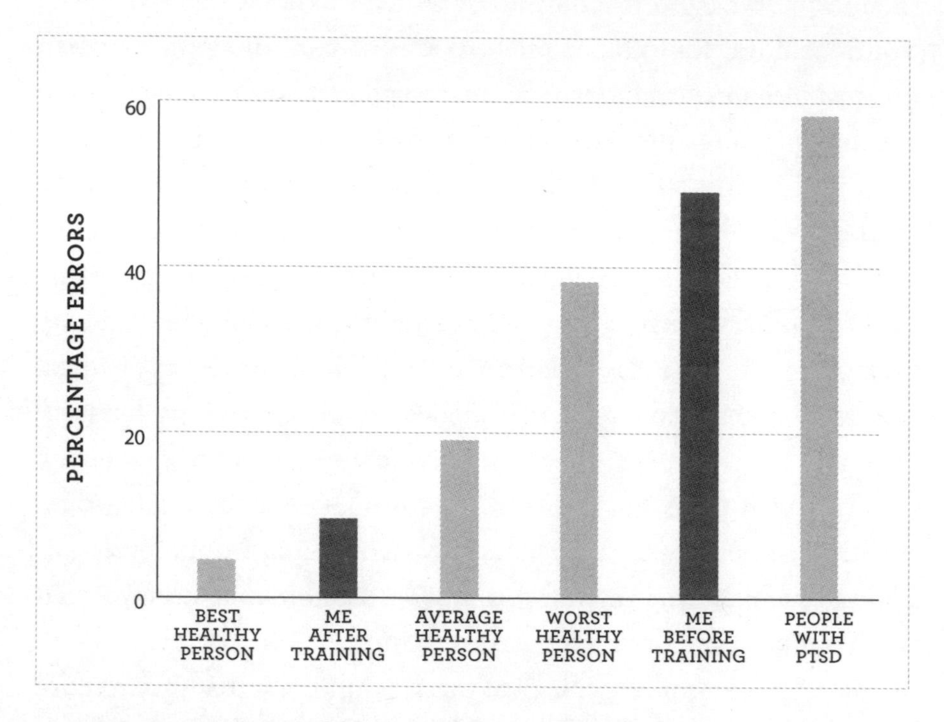


**Fig 4. Percentage errors in the 'Don't Touch Betty' test.**

Joe was as amazed as I was. 'We were like, "What? Did we run the same version of the test? That's remarkable,"' he said. And in fact, I asked them the same question straight after doing the test, because the experience of doing it was so totally different. This time, it felt like I had all the time in the world to spot Betty — and when she came on the screen she didn't smirk before disappearing; instead, she gave me a friendly smile as if to say 'Hi', before slowly fading away. A few times, I even smiled back. Everything felt like it was happening in slow motion. But they checked, and it was indeed exactly the same test, and strangely the time it took me to press the button was exactly the same before and after the training. Real time hadn't changed, just my perception of it.

But might I have simply become better at controlling my hand to stop myself from pressing the button? The training is similar to the Betty test, after all. I put this to Joe — but he tells me that, on the basis of their other studies using this training, they would expect up to a 4 per cent improvement from practice alone; I improved by more than ten times that. Moreover, my accuracy on the blink test — measuring how soon the brain can refocus after a distraction — showed similar improvements, scoring 46 per cent before training and 87 per cent afterwards. 'That's a huge improvement,' says Joe, and the blink test not only doesn't involve pressing buttons at all, but has even less of a practice-effect than the Betty test.

Okay, so if the results are real, what have they done to me? Have I really changed my brain in just one week, after only four hours of, boring but fairly easy, brain training and a side-order of stimulation?

'Not structurally,' say Mike and Joe in unison, keen to rid me of any notions that they have rebuilt my brain. 'But functionally, how you engage the brain ... something is different,' says Mike. That means that, while I might not have wired in a completely new circuit, the one I already have might be working more efficiently.

In a way, that's even more exciting, because it means that you don't have to make huge structural changes to the brain to fundamentally change the way it works and — crucially — the way you experience life. Just a bit of effort in the right direction can seemingly make all the difference.

It does mean that it's not easy to tell exactly what changed in my brain because we didn't do the training in the brain scanner or do a before-and-after MRI scan. This isn't because they forgot: MRI provides only a broad-brushstrokes picture of the brain and, after only a few days, such changes, at the level of a few connections here and there, would be too small to show up. So we have to extrapolate from other studies, and some of the psychological tests they asked me to do earlier in the week. Which is actually how science often works: comparing new results to an existing body of knowledge to work out what might have happened.

It seems that the changes Joe has measured, and I have felt, come down to the way the training has been designed. The first thing is that it is no accident that the training is boring — it has to be, otherwise it wouldn't tax your attention. In one pilot study where another group of researchers tried to make the training more exciting for children, all the benefits of the boring version promptly disappeared. Joe tells me that the boringness engages something called 'tonic attention', which is best described as an ongoing 'ready' state where you are watching to see when the target will pop up, while the unpredictable nature of the targets taxes 'phasic attention', or moment-to-moment fluctuations in alertness. The best state of mind for catching all the target images is one in which you engage tonic attention just enough to keep track of things, while being alert enough to react when a target pops up.

In other words, you have to be in 'the zone': that elusive sweet spot where everything feels not too difficult, yet challenging enough. That's why Mike and Joe were looking for a 50 per cent–correct score to build on in the training: this is each person's individual 'zone', and, once they've found it, they can start to ramp up the difficulty and gradually build up the skills. It took them a while to find my zone, but once they had, half-way through day three, I was able to build on it, making it the go-to state that my brain used whenever I needed to pay attention.

It certainly feels like I've accessed this state now. But what is 'the zone' exactly? Neuroscientists have been trying to answer this for years, with various levels of success. But Mike believes it is a state of attention where there is perfect balance between the dorsal attention network and another circuit, called the default-mode network. These are bits of the brain that are activated when we are thinking creatively, or mind-wandering, or not thinking about anything in particular.

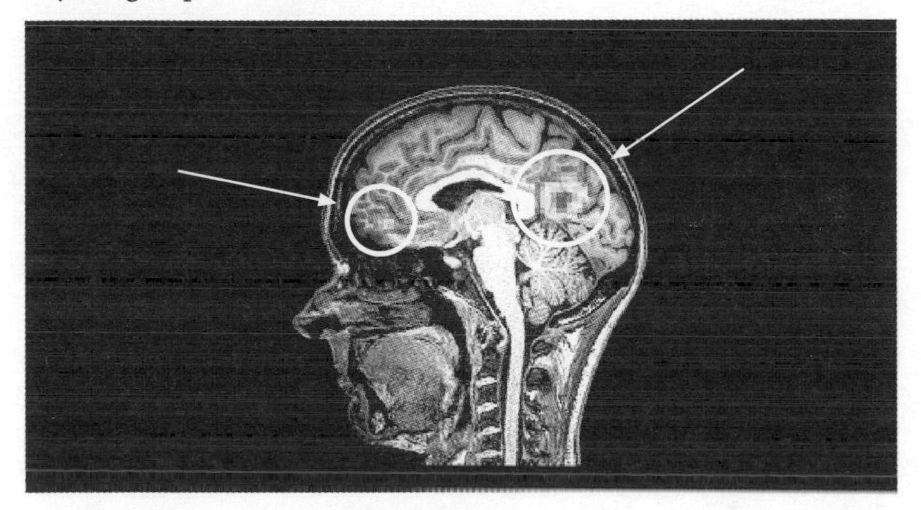

Fig 5. My brain, and the 'mind-wandering' circuit.

In a recent set of experiments, the Boston team found that when mind-wandering activity was at its highest, people were far more likely to make an error on a Betty-style test; more activity in the attention network proved better. Too much of either, though, backfired, and people were unable to stay on task for long periods.[2]

So the ideal state to aim for is not one with this mind-wandering mode turned off completely. It must have evolved for a reason — most probably for hunting and gathering purposes; a state of mind that allows you to scan the surroundings, waiting for something interesting to crop up, is useful in an unpredictable environment. Nowadays, it serves us well when we are casting around for ideas or just need a mental break. On the other hand, focusing hard is too exhausting to keep up for long, so the best way to sustain focus over minutes to hours is to turn it down a little bit — let the mind wander when it wants to, but not too far before bringing it back.

Because I didn't do the tests in the MRI scanner, there's no way of seeing what was actually happening in my brain, but a more consistent reaction speed on the Betty test also proved to be a sign of being in the zone. And I became way more consistent on the Betty test after my week in Boston.

Perhaps I started using the same basic circuits more efficiently, by engaging the right side of the network more, or by doing a better job of nipping my wandering mind in the bud. With the laws of brain plasticity being what they are, over time this might add up to larger areas of brain tissue and more connections between different parts of the dorsal attention network. Eventually, it might become part of who I am, just like my butterfly brain is now.

'Is that it?' I asked. Could focusing better be a simple matter of getting into a relaxed state of mind, concentrating, and letting the mind wander occasionally? It seems so. 'There are these fluctuations in performance and these go along with fluctuations in the brain. That may be a property of the brain — it needs to fluctuate, it's going to fluctuate ...' Mike begins, and then Joe jumps in, excitedly, the way he does. 'Embrace the fluctuations!' he exclaims. 'These ultimate states of attention are flow — you're on the ocean and you're riding into it and you're engaged ...' 'And it's not unpleasant,' Mike jumps back. 'That's key. It's hard, but it's not bothering you.'

Going with the flow has never been my natural state, and Tim Pychyl, a psychologist at Carleton University, in Canada, and author of *Solving the Procrastination Puzzle*, has some suggestions as to why.

He says that failing to concentrate is a basic flaw of the human psyche that we are all prone to, to a certain extent, but that some personality traits make it more likely than others. Three traits in particular hijack focus: impulsivity, anxiety, and a lack of conscientiousness. In tests I have done so far, I score high on two out of the three.

First, the State-Trait Anxiety Inventory (STAI-T) measures how prone you are to (usually pointless) worrying. When Mike gives me scores on this measure, he starts to looks a little uncomfortable. 'So, errr ... this isn't a clinical measure ... but, you did happen to score relatively high on trait anxiety.' I'm actually in the 86th percentile — another way of saying that, out of 100 people, 86 of them are less anxious than me.

Mike's hesitance to reveal this particular result makes me think that I am probably on the cusp of some kind of anxiety disorder. Given the prevalence of neuroses in my family, that wouldn't be much of a surprise. But I feel like I should justify myself, and I point out that if you're not anxious then you probably haven't understood the seriousness of the situation. 'Absolutely,' says Joe. 'If you're not worried then you're probably in a job that will be replaced by robots!'

Perhaps. But whether you call it anxiety or realism, it certainly doesn't help the brain to focus — which makes a lot of sense in evolutionary terms: if you're trapped in a cave, with lions closing in from every side, now is probably not the time to sit down and focus on knapping the perfect spear tip. The brain has a mechanism to widen focus when danger is near. Likewise, getting stressed about not being able to concentrate only releases a flood of stress hormones into the brain, and they don't help in the slightest. 'When you're not too anxious and you're not too engaged and you're kind of in this sweet spot, these certain receptors in the prefrontal cortex called the alpha-2A receptors are on. And if you get too stressed, they shut off,' Joe says.

So the key to focusing better is to relax and not try so hard. Absolutely, says Joe: 'If you're always on then you can grind yourself into a little nub, and you're fighting yourself the whole time and you do worse work.'

On top of that, being worried also takes up brainpower that could otherwise be used to stop the brain's impulsive tendencies to skip off and think about something else. I scored high on the standardised test of impulsivity, too.

Low conscientiousness, though, doesn't sound like me — I've always been a massive swot — and, sure enough, I score pretty high on the conscientiousness scale. But might this also be part of the problem? What if my impulsive streak is constantly trying to take me off task, and my conscientious streak is desperate to stay on track? Is there some kind of mental tug-of-war between these two traits? If so, given my anxious temperament, this is only going to cause worry and stress, and shut off those all-important receptors that I need in order to be able to focus.

What a mess. But at least my week with Joe and Mike seems to have sorted that out. Joe whips out a graph of error rates across all my sessions. 'In the beginning, it was 90 per cent, and then we were playing with things and we were trying to get you below 50 per cent. You can see this big drop here, and then another big drop, and then the last couple of days you were just kicking butt. I was like "wow", and everyone in the lab was like "What? That's not possible,"' he admits.

In fact, he tells me, even compared to the PTSD and stroke patients they have worked with, it was a huge improvement. 'You're going from basically making a mistake 90 per cent of the time to like, "Oh, I've totally got this."' Like doing progressively more sit-ups each day to strengthen stomach muscles, the training does seem to have strengthened the brain circuits behind a 'relaxed and ready' or 'flow' state of mind and made them work more efficiently. And, because the brain always chooses the well-trodden path over a less well-used network, being in the zone should soon become my default state of mind when I sit down to work.

Then Joe gives me the bad news. My newfound calm almost certainly won't last unless I do something to keep it going. It's the downside of adult brain training, apparently. Just like physical exercise, you have to keep at it or you'll end up as flabby as before. And since my personality and brain are primed for mind-wandering, I will probably slip back towards the butterfly baseline where I started. 'The dose you had will probably last two or three weeks,' he says apologetically.

Now what? Joe promises to send me more training when I get home, but I can't expect him to do that forever, and I can't keep crossing the Atlantic for a top-up of brain stimulation. And there's no app I can take home with me either. Joe and Mike are more than happy to admit that they still don't exactly know what the training does to the brain — and indeed if it really does transfer to everyday life, especially in healthy people who manage to lead a normal life. In fact, they have never tried this on someone who didn't have serious problems with their brain before, and they are as interested as I am to see how I get on after I get home.

So, did I keep the calm, even when the demands of juggling a young child, work, and chores kicked back in? Actually, yes. For a few weeks, I felt as relaxed and focused as I had during my stay in Boston. Life was good — easy even. Things that previously would have sent my concentration running for the hills, like playing Lego with my son when I had a million jobs to do, were actually pretty enjoyable.

Although that feeling was never going to last, afterwards, I at least could remember what it felt like to be in the 'zone' that I had practised in Boston. And when I found myself not in the zone,

I could now at least notice it and try to do something about it. Thankfully, it turned out that there are other ways to get the same effect, and there are a few decent scientific studies that indicate how.

One of the best options is meditation. It shouldn't be that much of a surprise — after all, Buddhist monks have been working at sitting still and being calm for centuries. Part of the reason for my starting on this journey was that when I looked into how to apply what we know about plasticity to my own brain, the only practical advice I could find involved meditation. But meditation has always sounded a little too new-age for me — and, seriously, who has the time?

Sara Lazar, a neuroscientist at Harvard Medical School, thinks I should *make* the time, if only for ten minutes a day — but every day. Lazar has spent years studying the effects of meditation on the brain, and has found that long-term meditators have lower activity in a region of the brain called the posterior cingulate cortex, part of the default-mode network that controls mind-wandering. Even after an eight-week course of mindfulness meditation, she tells me, there are changes in the brains of total beginners.

Yoga is almost as good, apparently, and since I do that at least once a week, I am already at least some of the way there. 'People tell me they don't meditate, they do yoga — but yoga is really a moving meditation,' she says. Lazar suggests maybe doing 20 minutes of yoga every day instead. I manage it for about a week, and then the good intentions start to wear off.

There are other options, though, and at least one of them has proven much easier to keep to, because it involves almost no effort. In lab studies, people who have run out of steam on an attention task were able to rejuvenate their attention simply by spending a few minutes looking at a picture of a natural scene. Looking at city scenes had no such effect, which seems to suggest that there is something magical about looking at, or preferably spending time in, the great outdoors. Other studies show that it works even better if you exercise at the same time.

Like all that mirror writing at school, I may already have stumbled upon a solution, having recently taken on a restless sheepdog puppy who needs a lot of running around outside. Now, whenever my brain starts to misbehave, I down tools and take him for a stomp through the woods. So far, nine times out of ten it does the trick. After a long, muddy walk, I feel about as close to my Boston self as I have since coming home from that trip.

Ultimately, though — and I don't have any scientific data to back this up, but Joe agrees that it's plausible — anything that gets the brain into the same relaxed-and-ready state that I reached in training should have a similar effect. The key is to find something you like doing, then do it at a level that's just difficult enough that you have to concentrate but easy enough that you still enjoy the experience. For me, that's yoga or a long walk in the woods; for my husband, it's climbing up a high rock face. For others, it's singing in a choir, swimming in cold lakes, running marathons, or playing in a band. Whatever it is, my experience with the advantages of getting in the zone suggests that it's a good idea to do it as often as possible — and, with a bit of luck, that relaxed but focused state of mind will become a default zone, just waiting for when you need it (see Box 1 below: Attention-hacking for beginners).

For now, I would say that hacking my attention system is very much a work in progress. I still have butterfly days, but not nearly as many — and having scientific data that backs up the idea that I'll get more done if I take a break and go outside helps me stop and get serious about re-setting my brain's attention network. And at least I now know what the 'zone' feels like, and am more aware of when I'm in it — and importantly, when I'm so far out of it that I might as well abandon my desk for half an hour and go for a walk. More than that, though, I had set out to ask the question of whether I can use neuroscience to improve my focus, and the answer is a resounding *yes* — albeit in a slightly different way to how I had expected.

Mike and Joe don't seem to think that I have made any structural changes to the way my brain is wired. But, even so, I can feel that there has definitely been a change, more likely brought about by changing the way I am able to use what I already have. The zone has always been there, perhaps, it's just that I have never been able to control it before. If so, then my new, better-focused mind should be more capable of bringing about change in other areas of my brain.

The next step is to curb that anxious streak. After all, if my anxiousness is getting in the way of focus, it's getting in the way of everything that requires focus, too. If I can change that — and there are plenty of scientists doing research on anxiety — then maybe other cognitive tasks will get easier. And although I have a hunch that my anxious tendencies stem from a genetic glitch running down the female line of my family, in theory, genetics are nothing compared to the powers of brain plasticity. Or are they? It sounds like the perfect challenge to tackle next. Worry genes versus bloody-mindedness and brain plasticity? Bring it on.

### *ATTENTION-HACKING FOR BEGINNERS*

1. Stop stressing. Stress releases a flood of hormones into the brain that tells your brain to widen its focus to look for danger and then zooms in — probably on the wrong thing entirely.
2. Find something that you enjoy that requires focus, but is relaxing and enjoyable. It might be sport, craft, learning a language, cooking, chess — whatever. But find it ASAP. It gets your brain into the right zone.
3. Practise point 2. A lot. And never, ever, feel guilty about taking the time out to do it (see point 1). The zone needs practice so that it is strong enough when you really need it.
4. Aim to meditate for 40 minutes once a week and 10 minutes every other day. If (like me) you can't sit still, 'moving meditations' such as yoga, swimming, or walking might do the trick.

While you are waiting for your brain to change and make this all seem a little easier, there are a few strategies that have some scientific credibility to back them up. These include:

5. Make your task visually stimulating or add background noise to max out your perceptual system and leave mind-wandering behind.
6. Take regular breaks, preferably outdoors. These must be a total distraction from whatever you are doing — so something that is either mentally taxing or takes all of your focus. Loud music or strenuous exercise might be a good option.

## THE MEDITATION DIARIES: PART 1

What, you might ask, is my problem with meditation? Evidence has been mounting for some time that it is really good for the brain, so why don't I just breathe in, breathe out, and give it a go?

If I'm totally honest, my problem with meditation is not only that it involves so much sitting quietly, although that is clearly a problem for me. It's also the fact that when people who do a lot of meditation tell you about how great it is, they always do it with this ... *look*. It's the kind of look you often find on the faces of devout religious types who really hope you will see the light too. There's a slightly smug undercurrent that implies they really feel sorry for you not having found it yet, and for some reason this makes me incredibly angry. Perhaps it's because I pride myself on being a natural sceptic. I like to start from the basis that everything is rubbish, questioning everything until I'm sure I'm not being taken for a ride. So, while I like the sound of the brain benefits, if I start with that dreamy half-smile look, you have my permission to smack me, because I've clearly been brainwashed.

On the other hand, there is no denying that there is a huge amount of evidence that meditation provides just the right kinds of changes that would be helpful for anyone seeking better attentional control, and to rein in a panic-prone mind. Recent studies have found that mindfulness meditation changes the way that the default-mode (mind-wandering) network is wired in to the executive control networks, presumably doing something similar to what Joe's training did — or would do if I kept it up long term. That sounds like something worth pursuing.

Another recent study also found that strengthening the frontal control networks gave people more of a handle on their emotional responses, which made them better able to cope with stress. This in turn seemed to reduce certain immune molecules in the body that are in charge of inflammation — something that should be kept to a minimum, for both body and mind.[3]

So here I am, in a church hall in my hometown with seven strangers (thankfully, since I was dreading bumping into someone I knew) and Gill Johnson, a meditation teacher who has trained with the mindfulness guru Jon Kabat-Zinn. Kabat-Zinn was a medic at the University of Massachusetts Medical Center when he brought mindfulness to the masses by rebranding it as a stress-reduction technique in the late 1970s. Since then, mindfulness-based stress reduction (MBSR) has become a global brand, and if you think you don't know anyone who does it, you probably do but they are keeping it quiet. So here goes.

Thankfully, Gill isn't afflicted by the 'look'. She is warm and down-to-earth and undoubtedly very chilled. In place of the look is an unwavering gaze that makes me feel that she knows all about me already. And she can rock a jumpsuit, cardigan, and slippers like no one I have ever met. 'This is meditation — we need to be comfortable,' she says with a grin.

As we wait for a few latecomers, Gill says we should 'sit with how that feels'. I think 'sit with how it feels' might be a phrase she uses a lot, and I think I might find it a bit annoying. What if I don't want to sit with feeling impatient? What if I want to huff and puff and fidget while getting more and more annoyed? Actually, I wasn't annoyed about the latecomers but was feeling very uncomfortable about the whole 'being in a room full of strangers' thing. Feels a bit like a therapy group or something ...

Finally, we get going on the first exercise, which is clearly designed for the sceptics in the room. We are to close our eyes and imagine holding a lemon, how it feels in our hands. Then we are to bring it up to our noses and smell it (am I imagining it or did I just smell lemon?). Then, finally, we take a big bite out of it, the juice running all over our chins and hands. Instantly, I'm salivating. It's quite a cool demonstration — I had expected to *imagine* tasting it, but not for my body to react as if I had. That's obviously the point: your mind makes things happen in your body, so be careful what you let your mind do ...

After a few more exercises, including mindfully eating a raisin (raisins taste like grapes when you stop to chew them — who knew?), we end the session with a body scan. This involves checking in with various body parts while lying on a mat on the floor. I spend most of the 30 minutes asleep, briefly wake up to realise that I should be focusing on a certain bit of my body, and that I feel cold lying on the floor, and then drop off again.

So far, I'm not sold. I find the session frustrating in the same way I found Pilates frustrating: you just don't seem to *do* very much. I walk home with a slight headache, and feeling lethargic. And I find myself in a bit of a bad mood for the rest of the day. I get virtually no work done — which is annoying, because part of the point of this is to help me focus. I've read the research, I know this is meant to be very good for my brain. I'm just not sure I like it.

CHAPTER 2

# Anxious All Areas

*'The assumptions you don't know you're making will only get you into trouble and confusion.'*

— DOUGLAS ADAMS

'I'd like you to think of something that you worry about often,' said Alex Temple McCune, a baby faced PhD student with the air of a much older, kindly doctor.

'Oh, that's easy. My son running into the road outside my house,' I reply. Alex stares at me, impassive. 'He's five,' I add, by way of explanation. 'It's a very busy road'. He nods, slowly. 'Okay, I am going to leave the room for five minutes and I'd like you to worry about that for the whole time I am gone. Think about it in as much detail as you can, and try not to think about anything else.'

Oh god.

'This is going to be horrible!' I plead, feeling my eyes widen, as Alex gets up to leave. He hesitates at the door, but goes out anyway, leaving me alone in a sparse, white, windowless room to think, in horrifying detail, about the worst thing that could happen in my life.

I am here at Oxford University to take part in a study into the cognitive basis of worrying, in the lab of Professor Elaine Fox. I've taken all the necessary screening tests to confirm that I am, indeed, a frequent worrier — and over the next two weeks the plan is to try and rectify that with a training course, designed to change the way the brain deals with stress.

After my success in Boston, changing this particular feature of my brain is very much next on the list. It seems to be one of the areas where there is pretty good evidence that you can, with a bit of effort, make real changes to your brain. There has been a good 15 years' worth of work in this area, because worrying too much is not only bad for sustaining focus, it is also seriously bad for your health.

Here's a statistic that worriers everywhere will enjoy: persistent worrying — even low-level fretting that doesn't qualify as a proper anxiety disorder — makes you 29 per cent more likely to die of a heart attack and 41 per cent more likely to die of cancer. In fact, according to the study of 8000 people that generated these figures, worrying a lot makes you more likely to die of anything, and the bigger the daily dose of stress, the greater the risk.[1]

I can't help thinking that if my friend Jolyon hadn't spent his twenties and thirties enjoying a totally debauched lifestyle, he'd probably live forever. Because Jolyon never, ever, worries. While I have been getting my knickers in a twist over various incarnations of this book, he has formed a company, launched two successful brands, and made several million pounds. He named the company 'Gusto' for good reason — this is a man who never does anything by halves, and couldn't care less what anyone thinks about him while he's doing it. You could say that he doesn't even worry when he probably *should*: he launched Gusto while his partner was pregnant with their first child, giving up a highly paid, steady job that he was very good at, and putting all their financial security on the line to do so. If it failed, as he cheerfully told me most businesses do, they would find themselves broke, with a newborn and nowhere to live. It was touch-and-go during the first few months, but he never really believed that it wouldn't work. 'I lost thirty-thousand pounds on one day when I first started,' he told me. 'It just made me more determined to make it back.'

It's the same story you hear over and over again about plucky overachievers: they don't get bogged down by what could go wrong or what just did go wrong, they just dig in and get on with it. It must be a really nice way to live.

Then there is the fact that anxiety is really bad for pretty much any kind of thinking. It not only narrows focus, it reduces impulse control, and robs the brain of a processing power that could better be used for other things. Over time, it has been found to shrink the hippocampus, a crucial brain area for memory. There has been some evidence unearthed recently that having an anxious temperament has some benefits, such as making a person more empathic and quicker to respond in a crisis — but overall it's not a great state to be in if you want to get the best out of your brain.

Some scientists — including Elaine Fox and her team at Oxford — think that differences between the likes of Jolyon and the likes of me come down to basic differences in the way our brains process information about the world around us. In her research, and in her book *Rainy Brain, Sunny Brain,* Fox argues that it all comes down to balance between two of the most ancient and powerful circuits in the brain — one responsible for seeking out danger and the other for spotting potential rewards — and how well connected they are to the newer, thinking bits of the brain. A skew to one way or the other is known as a cognitive bias — in other words, an assumption we don't know we are making. The direction and strength of these cognitive biases, Fox says, make us who we are — whether that be a driven and confident risk-taker like Jolyon, or more of a reticent worrier like me.

So again it all seems to come down to what we do with our limited budget of attention. Unlike the deliberate focus I was working on in Boston, though, this is an automatic kind of attention that operates within milliseconds, directing your focus to whichever parts of your surroundings seem particularly important. Crucially, all of this happens before we are consciously aware of noticing anything, and that means that, although we don't realise it, our conscious mind is constantly being fed a fundamentally skewed view of the world. This makes it particularly challenging to control. How can you change something that you're not even aware you are doing?

A negative cognitive bias might not be very good for you, but it undoubtedly evolved for good reason: it came in handy in the days when we were at the mercy of large toothy predators and men with clubs because it drastically cuts down processing time in the brain when we need to move fast. The downside is that the unconscious nature of these biases means that we live under the illusion that our own impression of the world — whether it is fundamentally safe or to be fretted about at every opportunity — is a totally accurate window on reality, when it is actually anything but. And that means that if you want to change your outlook on life — say, if you don't like the idea of hurtling, grey-haired and haggard, to an early grave — it isn't an easy thing to do.

On the plus side, the laws of neuroplasticity aren't swayed by a little thing like the border between unconscious and conscious processing, and Fox and others are working on finding ways to nudge troublesome cognitive biases back in a more positive direction. It sounds like something that is definitely worth a try, especially because some research suggests that all you need to do to retrain your cognitive bias towards a rosy view of life is to play a computer game for a few minutes every day.

It's a controversial area of research, and not everyone is convinced that it works, but it resonates with me partly because it treats an anxious temperament not as a fundamental part of who you are but as a kind of system error in the brain. And I get that, because if I'm really honest about it, my main problem with all this worrying is that it really isn't *me*. Outwardly, I'm quite a risk-taker (freelance journalism isn't for softies) and most people would probably describe me as mostly upbeat. The other day, another mum at the school gate described me as a 'super mum', and I don't even think she was being sarcastic. So clearly I give off the impression that I have things pretty much under control. Only I know about all the negative chuntering, worrying, and general unease that goes on under the surface and that, frankly, gets on my nerves.

Earlier, in Boston, I found out that I score highly on a measure of trait anxiety (also known as neuroticism, which seems a bit harsh) and research has shown that people like me, who score high on this measure, tend to have a negative cognitive bias, which we use to subconsciously scan the environment for threats at all times. We are also more likely to get stuck in threat-obsessing mode, assessing and re-assessing a situation in pessimistic terms, and getting progressively more worried. And yes, I do both of these things. The first of them, threat-readiness, pretty understandably, I think. When I was 19, my father was killed in a car accident. In the 20-odd years since then, I have perfected a kind of 360-degree, all-seeing eye for danger — especially the randomly occurring kind that could take a loved one away from me at any moment.

My age when my father died might explain why this harsh life lesson stuck fast in my brain. It has long been suspected that the adolescent brain is particularly plastic. This, after all, is a time when growing independence relies on being able to learn from your mistakes. Research has shown that the adolescent brain not only stores memories more vividly than adults, but is also particularly sensitive to stress, taking longer to recover from emotional setbacks.[2] These two factors together explain perfectly why unpredictable danger would have been writ large in my brain from there on in, but even in my most angst-ridden moments I know that the panic I feel is wildly out of proportion to any real threat. Is it helpful that if my husband is away on business and doesn't text me the second I expect his plane to land I start checking the news for air disasters? Is it good for my son that I jump out of my skin when he goes anywhere near a road, or even looks in the general direction of a hanging window-blind cord? And then I worry that his mother jumping in the air all the time actually makes it *more* likely that he'll have some kind of tragic accident — what if I accidentally knock him into the road while rushing to keep him from the edge?

Then there's the way that I assume the worst based on very little evidence. In the month before I went to Oxford, I sent Elaine Fox several emails explaining this whole project and asking if she would like to be involved in my mission to rewire my brain. I sent her the blurb for the book and a link to one of my previous articles, which was along similar lines to what I was hoping to do with her. And ... nothing.

The logical explanation is that she was busy. But, in my head, it was much more likely that: a) she thinks I'm some flake that can be ignored because the book sounds like trash and no one will read it anyway; b) she has read some of my previous articles and thinks I am the worst science journalist in the world; c) she is so disdainful of the whole thing that she has circulated my email to her fellow researchers and is now laughing heartily with them about this stupid journalist who won't leave her alone.

And how did I deal with this? Well, I put her number in my

phone and chickened out twice about calling her (because she obviously doesn't want to talk to me). Followed her on Twitter, just in case she's tweeting about the latest offering from the neuro-bollocks school of science writing. And I went around in endless emotional circles ranging from anger (it's a bit rude to just ignore my messages!), to indignant imaginary conversations where I justify my credentials to her, to practising how cool I would be about it when she tells me to get lost. Seriously. It was starting to feel like the time I phoned a boy in the year above me at school (also called Fox, funnily enough) to ask him out. I did it in the end, but not before practically giving myself a coronary. (He had a girlfriend, but we became good friends and briefly dated a few years later ...)

And guess what? When I finally did call Elaine Fox, she was lovely. She apologised for not getting back to me, and explained that she had been buried under an awful deadline for weeks and was still up to her eyeballs in it, preparing to move her entire lab over the summer. But she thought my project sounded fascinating and agreed to talk properly in a week or so. So what was the point of all that stressing? Weeks spent swirling around in an emotional whirlpool when I could have been doing something useful, like writing the introduction to this book.

This is the kind of thing I do to myself all the time and, while it has never really stopped me from pursuing what I want in life, it would be a lot more convenient to bypass the terror and self-flagellation and just get on with it. And, let's face it, it's ridiculous to get uptight about contacting a person who works on treatments for anxiety.

Of course, it was possible that my neurotic tendencies would have nothing to do with a wonky cognitive bias, so while I was waiting for her to reply, I headed to Elaine Fox's website[3] where there are two tests — one for cognitive bias and a questionnaire to measure a tendency towards optimism or pessimism. Just for fun, I asked Jolyon to do it too. Our scores are below.

| | Pessimism (Out of 24, higher score = more optimistic) | Cognitive bias (On a scale from -100 to +100. 0 is neutral) |
|---|---|---|
| Me | 8 | -31 |
| Jolyon | 21 | 51 |

Table 1. Optimism/pessimism scores and cognitive bias, as measured on www.rainybrainsunnybrain.com

Psychologists measure cognitive biases using a computer-based puzzle called a 'dot-probe task'. First, a cross appears in the centre of the screen, to give you something to look at. Then, two images flash up for 500 milliseconds, followed swiftly by a target (which can be anything — an arrow, a dot, whatever). Your job is to press a button on the left or right, depending on where the target (aka probe) has shown up. Research has shown not only that, a) people with an anxious temperament are quicker to spot targets that appear on the same side as the angry face (a negative bias); but also that, b) people with a negative bias are more prone to anxiety disorders and depression.

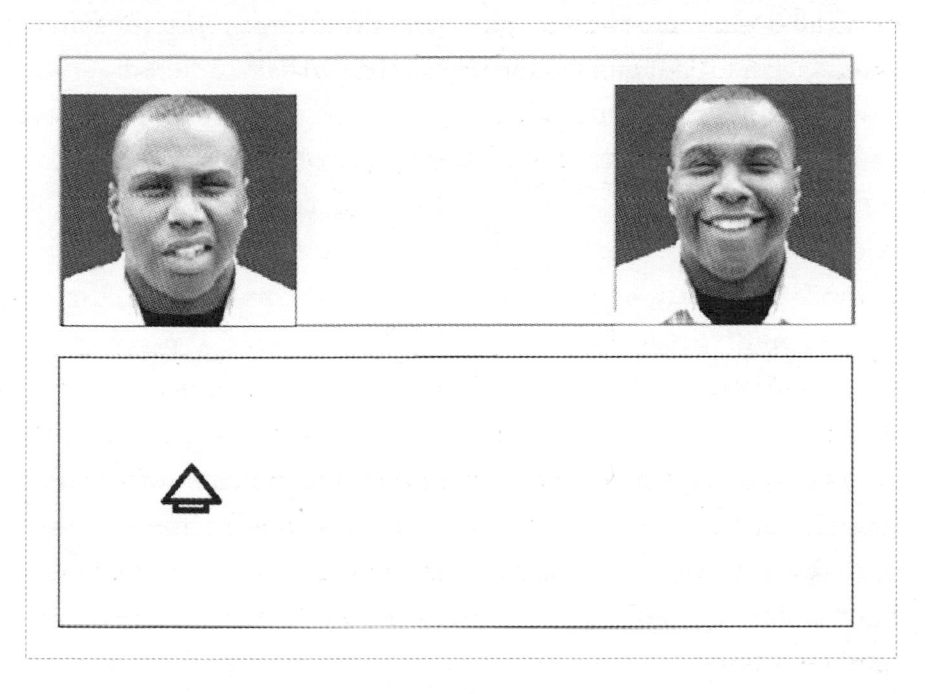

Fig 6. Dot-probe task screens 2 (top) and 3. (Courtesy of Mark Baldwin, McGill University)

A little bit of background research later and I have the proof: Jolyon is as strange as I am. According to surveys of large numbers of people, an average score on the optimism/pessimism test is 15 out of 24, meaning that the average person is generally slightly optimistic. Jolyon and I are each six points away from the average, but in opposite directions. He is unusually positive (hence the risky financial deals) while I am an almost perfect-scoring pessimist.

Which ties in nicely with the direction of our unconscious cognitive biases. A score of -31 means that I press a button 31 milliseconds faster when it follows an angry face than when it follows a smiley one. Jolyon, on the other hand, is 51 milliseconds faster to spot the target when it comes after a happy face. His brain automatically seeks out the good side of life — which probably explains his unusually high levels of optimism.

If all it takes to put things right is a quick bit of computer-based training, then it shouldn't matter how we got to be so different — but I must admit I'm curious. A glance around my close relatives makes me suspect that at least some of my neurotic tendencies might be inherited. On one side of my family it's considered unusual if you're not anxious, prone to depression, or emotionally a bit volatile. A quick headcount of my aunts, uncles, and cousins on that side of the family reveals that while the average number of people with emotional issues in the UK is one in every three, our family has roughly twice as many.

Elaine says she'd be happy to put me into her next study on the genetics of worry but they probably won't be ready to do the tests for at least another year. As a backup, she puts me in touch with the lab that does the analysis for her studies. She warns me that it'll be expensive (at least £500), but it might be possible as a one-off. It's my only hope, because while some medical insurance companies in the US will do the test under certain circumstances, commercial gene-testing companies like 23andMe — which offer off-the-shelf tests for risk genes, for everything from Alzheimer's to male pattern baldness — don't offer it, at least not yet.

As it happens, when I contact the lab, they are happy to help and can even chuck a few more samples in for me for no extra cost. I get a testing kit sent to Jolyon, we both swab our cheeks, pop them in the post to the lab, and wait with bated breath to see what comes back. After several weeks of back and forth between Jolyon and the lab, where he manages to avoid getting any DNA on his first set of swabs and the second batch get lost in the post — and I start to wonder how he runs a successful business at all — we finally get the results. And the verdict is ... not at all what I expected.

## THE WORRIER/WARRIOR GENE

Let's not get carried away — there are most definitely lots of genes that contribute to something as complex as your outlook on life. But the serotonin transporter gene is particularly interesting because of the way that it affects the kind of implicit learning that underpins unconscious biases.

The serotonin transporter gene makes proteins that mop up one of the brain's chemical messengers after they have been used to send a message between neurons, and recycles as much as possible, to be used again.

Everyone has two copies (aka alleles) of this gene, and it comes in two versions: one short (the S-version) and one long (the L-version). Confusingly, there are also two versions of the L-version ($L_A$ and $L_G$) and $L_G$ acts almost exactly like the S-version.

This matters because the S and $L_G$ versions contain less DNA and so they can't produce as many transporter proteins. This leaves more serotonin hanging around between neurons, meaning there is less serotonin in the actual neuron to send more messages.

Each of us has one of the following versions of the serotonin transporter gene: SS (short), $SL_A$ (medium), $SL_G$ (short), $L_AL_A$ (long), or $L_GL_G$ (short), with SL being the most common and SS the rarest, at least in a UK population (it varies around the world).[4]

People with at least one S or $L_G$ allele have been found to have a more reactive amygdala (involved in danger detection) than others.[5] They are also more likely to have a negative cognitive bias, to have a higher score on standardised anxiety tests, and to be at higher risk of depression. Other studies have found that people with the long version are more likely to have a positive bias.

But while it sounds like it's 'case closed' for the 'worry gene', it turns out to be not quite as simple as that. While some studies have found a link between the short allele and an anxious temperament, others have found the opposite. A long-term study led by Avshalom Caspi and Terrie Moffitt in 2003 provided a crucial clue as to why. They looked at not only genetics but also at the number of stressful events a person had been exposed to. And they found that people with the short version were indeed more likely to become anxious or depressed, but only if they had also experienced at least one stressful life event (divorce, abuse, death of a loved one). Without life stress, they were actually *less* likely to become depressed than people with the long gene-type. So the same gene can be either a 'worrier gene' or a 'warrior gene' depending on what happens to you along the way.

One possible explanation for this is that having the S-version gives a person a kind of super-plasticity brain, which learns and retains the lessons life throws at it unusually well. The downside is that if you learn from stressful experiences, your brain is more likely to conclude that there is a lot to fear in life and that the world is generally a dangerous place. On the upside, it quickly learns the opposite, too, if given half a chance.

Sure enough, in Elaine Fox's lab, people with the short version were quicker to develop not only negative biases but positive biases too: 'If a few bad events happen, you are more likely to develop a bias and then that gets reinforced. But by the same token, if the positive biases start developing, the people with that same genotype will be more likely to develop a positive mindset,' she told me.

So, the current consensus is that people with shorter versions of the serotonin transporter gene seem to be more vulnerable to the long-term effects of stress, but are also more likely to learn from life experience if it happens to be good. Given the way my brain has learned from my own stressful life events (my parents' divorce when I was five and my father's death when I was in my late teens), I would have bet money on me being a carrier of the short version of the gene. Jolyon, I wasn't sure about — but guessed that he was probably a super-resilient long-version carrier. We joked that if we were both short-gene carriers, it would be fantastic news for his parents and a bit of a kick in the teeth for mine.

In fact, the genetic test revealed that he is indeed an SS carrier — born with a brain that is set up to learn from life experience unusually well. The way he describes his early life suggests he got pretty lucky: no emotional upheaval, nobody died, and he wasn't bullied or abused in any way. This combination of genetics and experience seems to have made him super resilient to life's ups and downs, highly resistant to stress, and with an optimistic view on life.

Me, I'm among the 50 per cent or so that has one copy of the short version of the gene, and one copy of the standard long version. Which basically means that I have a moderately sensitive brain: not especially resilient, but not the most sensitive on the planet either.

This might be my negative bias talking, but it seems to me that it might give me the worst of all worlds. With one copy of the short type, it does mean that my genes have made it more likely that life stress will turn into a potentially problematic negative bias. But, not having the most plastic brain possible, it might not be an easy ride to change it, either.

## THE WAY OF THE WORRIER

At this point, it seems sensible to look at what happens in the brain to generate an anxious thought, to get an idea of what I would be changing, if I were to manage to turn things around. As it happens, this is one area where the circuitry is fairly well understood — at least in as much as anyone understands what a thought actually is. I once asked Geraint Rees, a prominent neuroscientist at University

College London, to explain what happens in the brain to make a thought, and, because he is a prominent neuroscientist, expected a concrete answer. What he actually said was: 'A thought is a mental state. It is widely believed that mental states (things in the mind) correlate with neural states (things in the brain), but the mapping between the two is not well known.'

So even top neuroscientists don't know how brain activity turns into conscious thoughts, anxious or otherwise. The tagline on Geraint Rees' lab website puts this rather nicely, I think, with a quote borrowed from Einstein: 'If we knew what we were doing, it wouldn't be called research, would it?'

But this we do know: for us to notice the thing to be feared at all, it has to stimulate one of our senses. Sensory information is constantly being passed to the thalamus, which sits in the middle of the brain and acts as a switchboard-cum-relay-centre between the senses, cortex, and other important areas like the amygdala and the hippocampus (which shunts things in and out of memory).

If there is anything around that has previously been marked as 'dangerous', it sends the information quick-smart to the amygdala, the brain's burglar alarm, which snaps attention to it while readying the body for 'fight, flight, or freeze' (racing heart, sweating palms, and so on). It also notifies the cortex (the thinking bit), which sets about making sense of the situation, and generates the conscious feeling of fear.

Joseph LeDoux, a neuroscientist at New York University, was among the first to describe this basic circuitry, and since then, an overactive amygdala has become the default popular explanation for everything from panic attacks to feeling a bit stressed out. It has never really seemed to explain my experience of being a chronic worrier, at least not for all forms of worrying. It probably accounts for the sudden tension I feel when crossing the road with my son, but when I'm fretting over something — like whether my work is up to scratch or whether the thing I said earlier made me come across a bit weird — I don't get a surge of adrenaline and a racing heart. It's very much a slow-burn, thinking-based, self-torture kind of reaction, not a primal, physical, need-to-get-out-of-here-right-now kind of feeling.

It turns out that LeDoux now spends a lot of his time trying to explain the subtle distinction between fear and anxiety — most recently in his book *Anxious*. He says that these are two different, although closely related, emotions and each has a slightly different bit of brain circuitry. Fear — the sweaty palm bit — is a physical reaction to a threat that is happening right now, in front of you, and that might kill you if you don't fight, run away, or hide. The *emotion* of fear comes later, once the slower thinking bit has come online. Anxiety and worry, on the other hand, are less to do with reacting to, or making sense of, a threat you can see, but are more to do with a feeling of uncertainty about if and when something bad will happen, and whether you are up to the challenge of dealing with it if it does.

Experiments have shown that damage to the amygdala doesn't prevent this kind of self-torture anxiety from happening. Instead, another brain area — the bed nucleus of the stria terminalis (BNST) — is in charge of processing uncertainty.[6] The key difference between them is where they get their information from. While the amygdala takes information mainly from the senses, to process what is out there in the world, the BNST gets more of its information from brain areas involved in memory and other types of cognitive processing, which means that it is more than capable of making a drama out of something that is entirely in your own head.

The BNST also seems to be the driver of hyper-vigilant threat monitoring: the keeping watch for anything that might, possibly, go horribly wrong. It does, though, put the brain in a state to react quickly if anything does happen. So the fact that I can freak out about my young son being near roads, even when he is safely tucked up in bed, should mean that I will react in super-quick time if anything actually happens. The downside is that I expend a lot of energy worrying about something that might never happen, and which my lightning-fast amygdala could probably take care of anyway if it did.

As it happens, the amygdala and the BNST are both connected to pretty much the same areas of the brain — most importantly, to the prefrontal cortex, which ultimately makes the decision about whether to escalate or calm down. The subjective feeling of anxiety or calm depends on which part of the circuit is most active at any point. Electrical activity is always flowing between the two, but depending on which is shouting the loudest at any one point, we will either feel in control of the situation or not.

If I want to take control of my worrying habit, then, I need to either reduce the number of (real or imagined) threats that get the amygdala and BNST going in the first place, and then use my thinking brain better to get out of the worry hole — or, preferably, both.

Gaining more control of the prefrontal cortex sounds familiar. But if Elaine Fox is right, there is a stage before this one that needs to be addressed: namely, the unconscious biases that are giving my prefrontal cortex more to do than is strictly necessary.

A trawl through the work of cognitive scientists working in this area reveals that there are three options. One is to use cognitive-bias modification training to train my attentional system to seek out positive things in an environment rather than negative ones. The most common approach to changing those biases involves flashing up a group of faces — most of them angry, and only one happy face — and asking people to click on the happy one as fast as they can (see figure 7).[7]

Fig 7. Cognitive-bias training. (Courtesy of Mark Baldwin, McGill University)

According to the people at the Baldwin Social Cognition Lab at McGill University who designed the task, the idea is that, 'Each time you drag your attention away from one of the frowning faces in the game in order to find a smiling, accepting face, this helps to build a mental habit. After a hundred trials or more the habit can become automatic ... the mental habit generalizes beyond the visual domain of disengaging from frowning faces, to apply to disengaging from thoughts and worries ...'

Another option, based more on the prefrontal cortex (that part of the brain above the eyebrows that keeps the more automatic impulses under control) is a different kind of cognitive-bias modification, but one that is focused on practising changing a negative interpretation of a situation. This generally comes in the form of audio of an imaginary scenario, after which you have to decide if everything is going to be okay or not. (You only get a point if you answer with a cheery '*yes!*'). Being forced to be positive over and over again is also meant to create a mental habit that gradually becomes engrained in the brain's wiring, blood vessel network, and so on.

Another option is to undergo working-memory training. Leaving aside the controversy over whether working-memory training does anything at all (that one is still being thrashed out), there are a few early studies that seem to suggest that training working memory might help to reduce anxiety — on the basis that improving working memory might provide more mental space to talk yourself out of a worry hole. Of course, it might backfire and provide more mental workspace to worry about *everything* that might go wrong.

Then there's meditation — which, for me, is a work in progress. If I'm totally honest, I'm really hoping that the answer to my brain blips isn't going to be meditation; I'm really not sure I can stick to a sitting-quietly habit for very long.

As it happened, my first bite of the cognitive brain-training cherry came not from Elaine Fox's group down the road in Oxford but from Ghent University in Belgium, where neuroscientist Ernst Koster is doing very similar work to Fox. While I am waiting for Fox to get back to me — and fretting about not having very much to write about yet — I drop Koster a line, to see if he might be able to help.

A couple of days later, we chat over Skype and he puts my fears to rest. Yes, he tells me, he could send me links to the online training they use in their studies, and he'd be happy to assess my anxiety levels before and after. It all sounds easy enough — and I wouldn't even have to leave my sofa. But then he tells me about the experiments they've been doing with eye-tracking technology: measuring attentional biases by monitoring where your eyes flick to, before you even realise they have moved. Apparently, Koster and his team have had some interesting results with this, but it can only be done in the lab — which would mean a trip to Ghent. I'm intrigued, and I've never been to Belgium ...

Koster promised to check the lab schedule and let me know when there would be some free time over the next few weeks. A few days later, I find myself on a Eurostar train to Belgium: he did get back to me as promised, but rather than, as I expected, setting a date for some leisurely time over the next month or so, I was informed that there was a rare gap in the lab schedule in just two days, and if it wasn't too short notice — and if I could get myself to Ghent — he'd be happy to put me through my paces.

It kind of was too short notice, but as if to prove that when there's no time for procrastination and worry I can just get on with stuff, I manage to book my travel and hotel, sort childcare and a dog-sitter, and pack a bag — in lightning-quick time. I arrive late in the evening and quickly decide against wandering the streets to find something interesting to write about my visit so far. What if I've accidentally booked into a hotel in the dodgy quarter? It was quite cheap ... So, with my threat-detection system very much in charge, I turn my back on the prospect of a cold Belgian beer and head upstairs to bed.

The next morning, Jonas Everaert, one of Koster's PhD students, meets me at the hotel and walks me through the backstreets of Ghent (the not-so-pretty end — maybe I made the right decision last night) to the University. He gallantly balances my bag on the back of his bike and we chat about all kinds of brain-changing technologies, including meditation. He tells me that brain-imaging studies with Buddhist monks have shown that their amygdala reactivity has dropped so low through constant meditation that it is no longer workable in modern life — they simply couldn't deal with the levels of stress and threat that the rest of us do. I don't want to take it that far, clearly, but a little more Zen might be nice. Still, it's interesting to get another insight into the pros and cons of meditation. The way most people are talking about meditation these days, you'd think it was the cure for everything.

Jonas takes me to Ernst's office — which looks pretty much like every scientist's office I have been to: a plain white room with piles of papers from floor to ceiling and all over the desk. While Jonas goes off to get us a coffee, Ernst welcomes me in and apologises in advance for the terrible coffee I am about to drink. He listens, seeming both interested and a little amused, to my bumbling explanation of what I'm trying to do to my brain. I can't help wondering if these researchers think I'm on some kind of fool's errand, and are politely humouring me because it makes a change from writing grant proposals. Does he really believe that he can turn my worry-prone mind into a chilled-out positive one in just a few weeks? I guess we'll see.

A few minutes later, there is a knock at the door and in come two of his research team. Ayse Berna Sari is shy and friendly, with the doe eyes of Audrey Hepburn and the big hair of Amy Winehouse. She and Alvaro Sanchez Lopez (who is a lot closer to the common stereotype of a scientist) point out to us both that it's going to be difficult for me to try their attentional-bias training programme, because it involves disentangling emotion-laden sentences — and they are all in Dutch (the main language spoken in this part of Belgium). Instead, they say, they can give me the working-memory training programme that Berna has been working on and see if it improves my performance on Alvaro's tests. In a recent study, they have found it to be useful in changing people's cognitive biases after just a couple of weeks of training five days per week. Despite everything I have already heard about working-memory training, and the controversy over whether it works, it all sounds very intriguing.

First come the now-familiar baseline tests, for which they take me to a white breeze-block-walled room with tiny windows too high to see out of. What is it with psychologists and their stark, windowless rooms? It's no wonder they do so well at finding anxiety in more or less normal people.

The first test was a more sophisticated version of the cognitive-bias test I did earlier, online. On the computer screen in front of me, a circle of black-and-white faces pop up briefly on the screen. Berna tells me they will either look angry, happy, or neutral. If all faces match, I am to do nothing. If one is different from the others, I press the spacebar.

It's actually surprisingly uncomfortable to have so many faces staring at you at once. The angry ones feel threatening, but, if anything, the neutral ones are worse — their blank stares make me wonder what they are thinking. I find this in real life, too. I'd rather someone was looking at me as if I was a piece of scum than just staring blankly. At least I'd know what I was dealing with. The smiling ones are kind of comforting, though — they give me the warm fuzzy feeling of being in a roomful of friends who like me the way I am. Consciously, I'd much rather look at the happy folk, but, like the online version I'd done, Berna's results show that it actually takes me longer — about 40 milliseconds longer — to drag my eyes away from angry faces and spot a happy one than it does to find a single angry person in a sea of smilers (see figure 7).

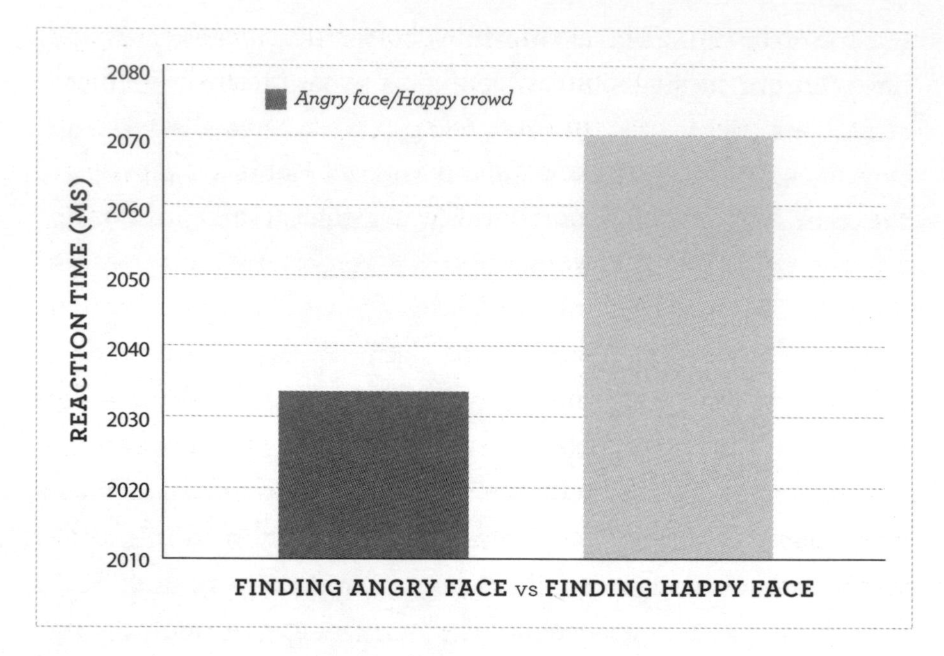


**Fig 8. It took me 40 milliseconds longer to look away from angry faces than happy ones, despite the angry ones making me feel uncomfortable.**

It doesn't sound like a lot, 31–40 milliseconds, but compared to participants in their previous experiments, it is quite high, Ernst tells me later. In a 2006 study, high trait-anxious people had a negative bias of 10–30 milliseconds, compared to less than 10 milliseconds in low-anxious folk.[8] Again, I find myself lingering near the end of the scale based on volunteers from previous studies. Oh dear.

After a few more baseline measurements — including one particularly horrible test where I have to look at photographs of sick babies and lonely old people, and then try to come up with a positive spin on the situation before rating how upset I feel — I go for a walk to rest my eyes. On the way back, I spot Alvaro in a room, setting up his eye-tracking experiments for me to try later. I persuade him to let me have a go at the eye-tracker now, and it's all good fun until Berna appears and gently points out that I'm in the middle of assessment and supposed to be taking a break. Fair point, but I make sure to get straight back to the eye-tracking room as soon as Berna has finished with me.

The last time I tried an eye-tracker in a psychology lab, about ten years ago, it looked like something out of Kubrick's *Clockwork Orange* — a pair of enormous glasses attached to a headset with cameras that stare into your eyes. But not anymore. Now, the equipment looks like a shiny pair of speakers attached to the bottom of a computer screen. Apparently, the shiny bit beams infrared light into the eyes, and a hidden camera in the screen is picking up the reflections from the pupils. Human eyes can't see infrared, so the only clue that the machine is following my eye movements is when Alvaro begins calibrating the machine and two white dots appear on the screen: the computer's-eye view of my pupils.

It looks as if an inquisitive little robot is looking back at me from the screen. When I blink, it blinks. When I tilt my head, it tilts its head, too. It's like having a virtual pet — I'm sure there's an app in this somewhere. It soon has all three of us in fits of giggles, until Alvaro snaps back into researcher mode, and instructs me to stay still so he can finish the calibration. I do as I'm told, and track a red dot around the screen while the eye-tracker follows my eyes.

First, Alvaro shows me their Dutch-language training test. Six words come up on the screen; I'm told each series of words makes up a jumbled-up emotional sentence, but only using five of the words. For example, 'I a person am useless worthwhile'. Depending on the direction of their bias, people tend to either make positive sentences ('I am a worthwhile person') or negative ones ('I am a useless person'). The eye-tracker can spot where your eyes first go to — so even if someone comes up with a positive sentence in the end, the computer knows you looked at the negative option first. Clever. As a training device, as the eye-tracker measures where your eyes flick to first, the words change colour: green for positive words and red for negative ones. Your aim is to go straight to the positive ones and avoid the red. I only know two phrases in Dutch, and both of them involve swearing, so clearly I have no chance of doing this test properly. After a couple of examples flash up on the screen, though, we both start to laugh. Even in a foreign language, my eyes seem drawn like a magnet to the negative words. Figures.

Next, Alvaro sends me back to the other room for a bout of working-memory training with Berna. I am to do two 20-minute sessions, and then do the horrible photos tests again — my least favourite so far — to see if anything has changed. I am sceptical that anything will change after just 40 minutes of training, but Berna tells me that they have seen a shift in experiments on larger numbers of people, so it isn't impossible. And, she says, the people with the most negative cognitive bias were the ones whose score improved the most after training.

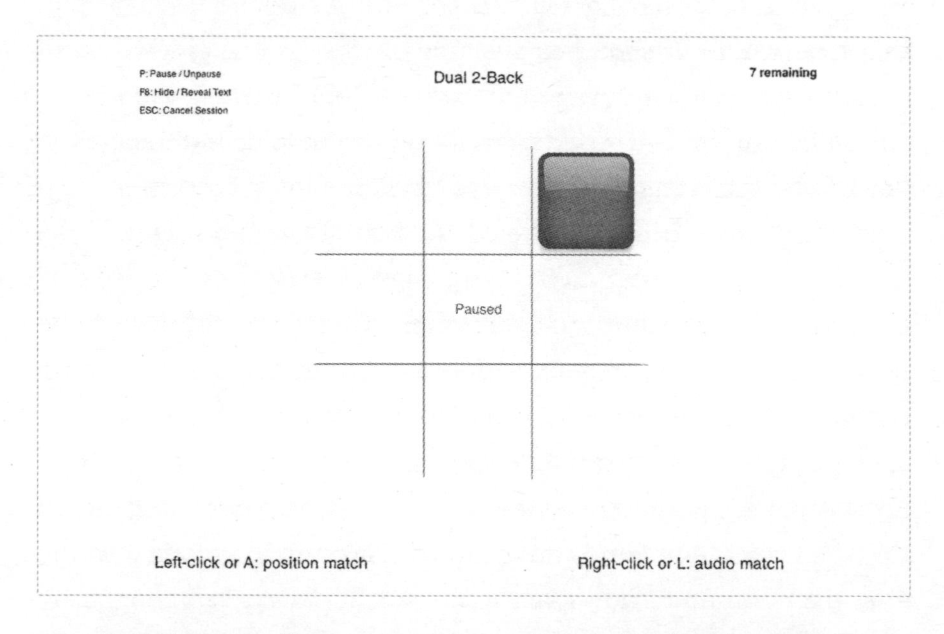


Fig 9. Dual N-back working-memory training. You can download the game at http://brainworkshop.sourceforge.net (Courtesy of Paul Hoskinson and Jonathan Toomim)

So, after two more really difficult 20-minute working-memory sessions, I again find myself looking at miserable photos. Depending on whether the word 'appraise' or 'reappraise' pops up on the screen afterwards, I am to spend the next 30 seconds dwelling on either the most miserable explanation or how there was a positive outcome in the end. When the picture showed a baby in an incubator, for example, I could either focus on the suffering of the baby and its parents, and the odds of it not surviving, or imagine that baby growing up big and strong and leading a happy and fulfilling life. Before and after each of my little reveries about the picture, I had to rate how I felt on a scale of 0 (alright, actually) to 9 (utterly miserable). I find this test pretty difficult; you only get a few seconds to think about each picture, which doesn't feel long enough to generate any real emotion — and the scoring seems rather fixed: who wouldn't rate their mood as improving after putting a positive spin on a sad picture? It all feels a bit forced. Still, I am intrigued to see if my scores have changed from before the working-memory training.

And according to this very quick bout of working-memory training, it did change my ability to compute different options (see figure 10). I even scored better than the average volunteer, for once.

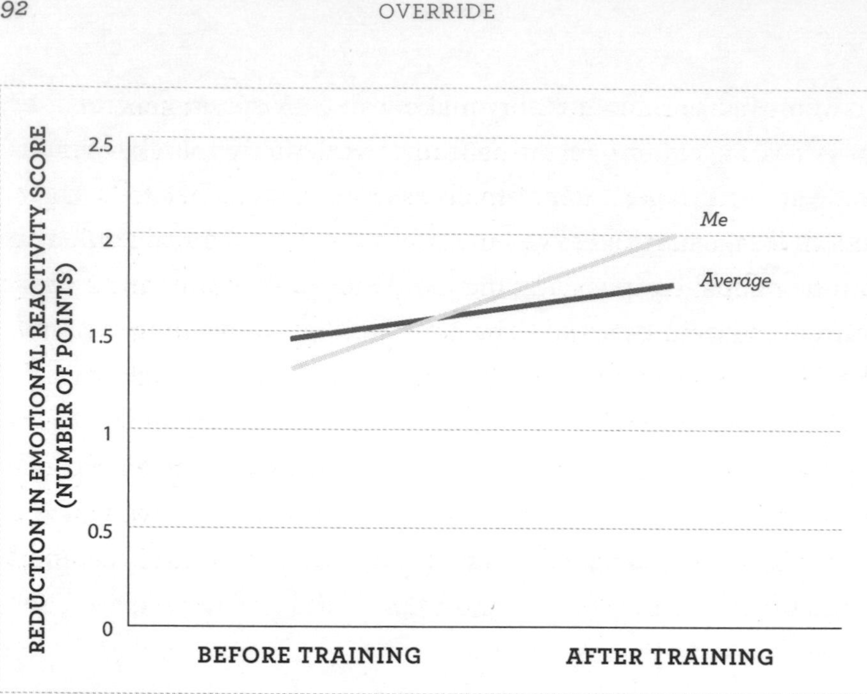


**Fig 10. I started off with a below-average ability to think positively about ambiguous photos but then, after 40 minutes working-memory training, shot ahead. The average is taken from the group's previous research.**

The idea behind improving working memory is that it buys room in the mind to weigh up different interpretations of a situation — to do the maths of: is it *really* all that bad or is it possible that I am overreacting?

I can't say I'm convinced that my results above reflect any real change in my thought processes in general, but at least if you take the scores with a huge pinch of salt (comparing one person against the average isn't what happens in science studies; they work by calculating averages from large numbers of people), it does look as if it may have briefly improved my ability to think a bit more positively. Would I do the training every day, though? It's quite boring ... Well, I'm about to find out, because I have promised to do their working-memory training for three weeks. Partly to see if it does anything for my anxious streak, and partly to test the brain-training spiel about an improved working memory making you smarter in general.

A few days after I get home from Ghent, though, I take a hiatus in my training when Elaine emails asking if I would like to take part in a training study in her lab in Oxford. Elaine's is a real study that will be published in a scientific journal at some point, and I don't want to mess up her results by doing two lots of training. So I put the Ghent experiment on the backburner while I go to Oxford.

The day before I am due at Elaine's lab, I get an email from Alex Temple-McCune telling me where to be and when. The email ends with a polite request to 'please be here on time'. By now, I've had quite a lot of experience with cognitive psychologists and I can't help wondering if this is some kind of trick to get me worrying for the purposes of the experiment.

They needn't have bothered. My travelling to Oxford coincides with a train strike, and the train I had been planning to get had been cancelled. The evening before I'm due in Oxford for the experiment, my anxiety levels go through the roof as I hastily make arrangements to drop my son off with a friend before school, so that I can get an earlier train. That night, I dream about empty train stations, being unable to find my car to drive to the station, and generally running around trying to get there on time.

It all works out fine and (despite the taxi driver triumphantly announcing 'here we are' on Keble Street, when I'd asked for South Parks Road, leaving me a good ten-minute walk from the lab) I get there bang on time and am met at the door by a smiling Alex.

First, I have to fill in multiple questionnaires about my state of mind today (state) and in general (trait). I have done a lot of these now and they don't get any more stimulating. Then I go on to do what I think is a working-memory test, involving coloured shapes in various positions on the screen, which I am supposed to remember having seen them for a fraction of a second. I do several sessions back to back (there are opportunities for short breaks but I'm not really sure where to look or what else to do — I'm in a white concrete box and it doesn't feel like the time for making polite chit chat).

Then Alex asks me to swap chairs so that we can do what sounds a lot like mindfulness meditation. My job is to focus on my breathing for five minutes and if my mind wanders try to bring it back to focusing on breathing. Randomly, the computer will beep and I have to tell Alex whether I am focused on breathing or on another thought. If it's another thought, I have to tell him if it is positive, negative, or neutral, along with a few words describing the thought. It's incredibly embarrassing. Examples of my thoughts include: 'This room is really cold-looking', 'I hope I don't fall asleep', 'I think I've got a headache coming', 'I hope I'm not ruining their experiment', 'This is taking ages'. All of them are negative. Some I don't even admit to. The thought, 'Has he noticed I missed my knees when I shaved my legs?' I kept to myself.

Then comes the 'five-minute worry'. This is as awful as I expected. I can see the car hitting my boy outside of my house and what it does to his head. I can see him lying there covered in blood and me screaming. I can imagine him being in the hospital in a coma, and the possible horrible outcomes — death, paralysis, brain damage. By the time they come back in, I am in a right state, feeling exhausted, tense, and downright miserable.

And now we are going to do the breathing exercise again. This time my thoughts start off even more negative: 'I want to go to sleep', 'I can see the horrible things I imagined in the five minutes', 'I've had enough of this'. Gradually, I go back to some kind of baseline, my last thought being: 'I feel more relaxed now than I did at the start'.

At the end of all this, I feel like a wrung-out rag and am happy when they demonstrate the training I'm to do and then let me go. The training will take about an hour a day, every day — even weekends — and I'm going to do it for ten days. The lab can track my progress online, so will know if I've been skiving. *Yep, okay, anything you say, just get me out of here.* Alex walks me around for a quick chat with Elaine and, still looking a bit concerned about my state of mind, says goodbye.

## POSITIVE-THINKING BRAINWASH GUINEA PIG: THE OXFORD STUDY

When I get home and start my training, it quickly becomes clear that I have been assigned to the control condition for the working memory part of the study. This I'm not happy about, but unfortunately this is the way randomised trials work: each subject is randomly assigned to a group whether they are student volunteers or nosy journalists wanting to have a go. I'm not supposed to know it's the control condition, but I happen to know from my time in Ghent that this kind of training should get harder as you improve — and it has been on the easiest level for ages, despite my near perfect scores.

I'm pretty sure I'm in the active (training) condition for the interpretive bias part of the study, though. The training is basically 20 minutes of audio of a lady with a lovely, gentle Dutch accent reading little stories about situations that you could view either negatively or positively. The situations all take the form of: '*So this happened, you freaked out about it, then decided it was probably fine. Is it fine?*' The correct answer, clearly, is 'Yes'. I get a green frame and a *ding!* for a correct answer and a red frame and a *buzz* for a wrong answer. It starts to feel slightly hypnotic, as if I'm being brainwashed into seeing the brighter side of situations.

Stranger still, when a stressful situation happens in real life, I find myself talking myself down in a calm Dutch accent. Okay, so there are only ten minutes before we have to leave for school and I still haven't made it into the shower. But my son is fed and in his uniform, his bag is packed and by the door, and I have cleared the kitchen after breakfast. I have just enough time to shower, dress, clean teeth, and go. Normally, I'd spend this last ten minutes of the morning running around, shouting at everyone and tripping over the dog. But today ... 'Is it going to be fine? *Yeeeessss.*'

It doesn't take long for the Zen to wear off, though. When the study is over, I try to invoke the calming voice of the Dutch lady whenever I worry, but, increasingly, it feels forced. Worse, Alex won't tell me whether my two-weeks of brainwashing made me any better at staying calm during the five-minute worry, because they will be analysing everyone's results together when they have collected all the data — another annoying feature of being in a real study, it turns out. But while it seems my ability to think positive probably did shift a bit, the effect barely lasted a week; so, even if it did give them an interesting result for their experiment, I don't think it is the cure I was hoping for. And if it takes 20 minutes of boring brainwashing every day to achieve, I don't think it would be a viable long-term strategy anyway.

So it's back to the working-memory training from Ghent, with a side order of online click-the-smiley-face — a bit of cognitive bias and a bit of increasing the mental workspace. I do the exercises religiously, every day for six weeks — even on holiday — and then take the cognitive bias and optimism tests again.

And ... something changes.

| | Pessimism (Out of 24, higher score = more optimistic) | Cognitive bias (Neutral = 0) |
|---|---|---|
| Me (before 6 weeks training) | 8 | -31 |
| Me (after 6 weeks training) | 12 | 35 |
| Jolyon | 21 | 51 |
| Me (after 15 weeks training) | 12 | 79 |
| Me (after 7 months training) | 13 | 93 |

Table 2. Before training, I had a resolutely negative bias compared to my optimistic friend Jolyon. After training, my cognitive bias shifted towards, and eventually past, his. My optimism score, however, remained stable.

I email Elaine the results. After pointing out that it's difficult to make real assessments from a sample of one, she did say this: 'Clearly your bias has shifted from negative to positive. This indicates that your attention is automatically now shifting towards the positive images, whereas previous it was automatically pulled by the negative — more similar to your optimistic friend. So all good and as expected.'

I keep the training up over the following weeks and months, and it goes even further — way past the giddy heights of Jolyon and heading towards 100 per cent. Interestingly, though, my natural pessimism score stays stubbornly where it began. I might worry less, but I don't expect everything to go my way.

Another measure of change is to re-do the standardised anxiety scale (STAI-T) that I did at the start of this process, in Ernst Koster's lab in Ghent. And again, it seems as if something has shifted. Before, my score on the trait-anxiety scale was 60/80. After, it was 49. 'There is a decrease, indeed,' wrote Koster, to my mind sounding a little unimpressed. Actually, it might be quite significant. According to research on large groups of people, a score of less than 48 indicates no anxiety at all, whereas 60 is pretty high — the kind of score you'd expect from someone with an anxiety disorder. Which is good news and bad, as far as I'm concerned. I started this process thinking I was just your average worrier. It looks like I started off with a big problem — but on the plus side, I have turned it around, and my score looks vaguely normal.[9] It's a success of sorts ...

It's nice to know the numbers are moving in the right direction, but it's more difficult to tell whether it has changed anything about the way I think in real life. And that's really the nub of it — it's no use changing your cognitive bias or STAI-T score if it doesn't make you feel any different. I can't ask anyone else if I seem more chilled — even my husband — because a lot of my worrying is a private affair in my own head (for what it's worth, I did ask him and he hasn't noticed any changes). Ultimately, a subjective take on whether life *feels* any easier is the real proof of the pudding.

The problem with this is that cognitive biases are under the radar of consciousness and by definition you can't be aware of the unconscious. Which means that if I have changed then I probably won't notice and if I think I do, I'm probably imagining it. Having said that, though, I do feel like I am slightly less self-conscious in situations where I have to chat to people I don't know very well. It feels like I might be getting better at spotting a friendly face and letting go of signs that may be interpreted as disapproval. This in turn doesn't give me any room to ponder whether that face is saying 'god, she's awful', or whether it bears no reflection on me at all.

Spending all that time looking at smiling and angry faces has also given me a handy demonstration of how a smile (or a grumpy face) can make other people feel. I find myself smiling at people more and saying 'how are you?' and that in turn makes them smile more at me — which is nice. This makes me think that it is probably worth keeping on with the happy-face clicking, to see if anything changes longer term. Nowadays, it only takes about five minutes, and I often do it while the kettle is boiling.

On the downside, since taking on the training, I have noticed an upsurge of anxiety-ridden dreams — ones where I have an exam and have forgotten to revise, where my teeth are crumbling in my head, or where my trousers are pulled down in public and everyone points and laughs at me. It makes me wonder if this is where my worries have gone to hide now that they aren't welcome in my conscious mind. Like that bit in *Eternal Sunshine of the Spotless Mind* where you find out that the memories are hiding so they won't be deleted, my mind isn't willing to give up its old ways just yet.

One thing I do know is that, looking back on the diary I've been keeping throughout the process, I don't feel like this anymore:

*Day 1 and 2 (July 2015)*

*I'm finding the training quite unsettling and it feels like it might actually be doing me harm. I'm looking at far more negative faces than positive ones and it's making me feel really uncomfortable. It takes ages to find the smiling ones and all the time I'm looking at the angry ones I'm getting more and more uptight. I get a wave of relief every time I find a smiling face. It's like a port in a storm.*

Now it's more like what I noted on my diary on day 63 (September 2015):

*... I don't feel like I did in the first week, where the angry faces were really upsetting. Now, it's just a quick 'bam, bam, bam' and it's done in five minutes. Really think I could keep this up long term ...*

As for the working-memory training, I'm less convinced. It proved difficult to stay motivated when my performance plateaued after several weeks. I carried on doing it for a while, but not for 20 minutes and not every day. A couple of months later, I stopped altogether. There just isn't enough evidence to convince me that it's worth my time yet.

The road outside my house still gives me the jitters, but I have found a way around it: we now go a longer but quieter way to school, riding our bikes along a wide pavement instead of walking up a very narrow one. I don't know why I didn't think of it before — maybe practising how to replace panic with a calm focus has helped make room in my brain to not just fret about the road but to think of a solution as well.

One thing that has proven particularly resistant to change: my ridiculous response to stress. About halfway through the Oxford study, while trying to set up the brain scans for the next stage of my mission, I had several knock-backs from researchers who I had hoped would let me have a go on some of their experiments. It was, various researchers told me, too expensive, and they couldn't spare the time to find a student to analyse my results, and that, anyway, I don't qualify as a subject for any study they were running. And even though I know that brain scans only tell you so much and that there are other, perhaps more reliable ways of measuring brain change, I did freak out on a similar scale to the Elaine Fox debacle when I was waiting to hear back from her (admittedly only for a few days this time, but I wasn't much good for anything during). I can't deny it was disappointing. I had hoped to knock this kind of thing on the head.

This goes back to the idea that to stand any chance of changing your brain, you have to pick your skills very carefully. An anxious temperament, I have found, isn't borne of one thing but many.

Social anxiety and performance anxiety are completely different things, and what helps with one may not help the other.

What I take from all of this is that getting on top of anxiety is, again, all about controlling attention. Unfortunately our attention is not always under our conscious control — and that's why self-help alone, at least in the form of advice such as 'think yourself better and you will get better' is never going to work. If you have a negative bias, you will always be swimming against the tide. It's why depressed people *can't* think themselves out of the hole: their brain is feeding them nothing *but* the hole.

Escaping the clutches of a negative bias is definitely not easy. I can't say for sure if cognitive-bias modification is the answer, and neither can Elaine Fox or anyone else. All I can say is that, for me, it seemed to help. But — and it's a big 'but' — 'pick your skills' definitely applies here. Train yourself to seek out happy faces, and that is what will change in the real world. I wondered at the start of all this whether fixing one neurosis would make all the others melt away too. The answer to that is a definite 'no'.

On the other hand, my mindfulness course is now well underway, and in theory I have a few more tools to counter a more general kind of angst. Mindfulness, I am discovering, is not just about the mind, it's about the body too. And that might be the key to greater control.

## THE MEDITATION DIARIES: PART 2

In *Fifty Shades of Grey*, Anastasia has a habit of biting her bottom lip whenever she's nervous. She doesn't realise she's doing it, but it drives the sexy billionaire Christian Grey mad with lust. Through the powers of meditation, I have discovered that I do something similar. Only I look more like Kermit the Frog:

**Fig 11. My worry face — or is it Kermit the Frog?**

Weirdly, I had never noticed this habit until Gill, my meditation teacher, gave us the session on noticing what is happening in the body; our homework was to pick a pleasant experience and pay attention to where we feel it in the body — and then do the same with an unpleasant experience. I noticed that, when I'm having a warm and fuzzy moment with my loved ones, I feel it as a warm spreading sensation in my belly — which is all very lovely, but not terribly surprising. More interestingly, when it comes to the negative stuff, I noticed that I have all manner of physical stress tics that come out whenever I get uptight. Kermit-face is the most common, but I also have a particularly unattractive frown, and a pursed-lip look like a disapproving granny.

The value of noticing this, Gill says, is that it opens up a gap between what the mind is doing and how it feels. Only then can we investigate what is going on. 'It's about noticing,' she says. 'Just noticing.'

The more I pay attention to these physical tics, the more I realise there is an underlying theme. I bite my lip when I am cleaning up the house, when I am thinking about what to write, when I am cooking, and when I am writing my to-do list. The underlying theme is: *I want to do this right. What if I screw it up? Am I crap at this?* Interesting.

Even better, now that I have noticed that I do this, I can consciously let go of my lip and point my attention directly at whatever it is I am so worried about. Then I can realise that either, a) there's no need to get all Kermit-faced about loading the dishwasher, or, b) I'm stressed because I care about the quality of my work — and that's a good thing. Either way, it helps to take the pressure off and give myself a bit of a break. It's actually pretty fantastic.

Another thing that I am starting to like about mindfulness is that it doesn't allow any room for giving yourself a hard time about doing things 'right'. Every bit of self-help that I have ever seen involves an element of 'You feel this way because you are doing it wrong. Do it this way and you will feel better'. Mindfulness allows you to notice your unhelpful habits without adding yet another nagging voice to the mix. Instead of 'I'm stressing; why am I stressing? I shouldn't be stressing, et cetera, et cetera,' with mindfulness it's more like, 'I'm stressing. I have noticed. That is all.'

I think I am starting to see a real benefit to this approach — it might be a way to reach the neuroses buried in my mind that cognitive-bias modification couldn't.

CHAPTER 3

# Let the Creativity Flow

Lila Chrysikou attaches the two electrodes on my head to a battery and slowly cranks up the power. As she does, my focus gets up and leaves the building.

It's a weird feeling. Until she flipped the switch, I was perfectly alert and on the ball. I'd arrived at her lab at Kansas University, in Lawrence, refreshed from a good night's sleep and two cups of tea, and was raring to go on a full day of experiments. We'd been chatting animatedly about my long journey to Kansas, the joys of parental sleep deprivation, and many things besides. But as soon as the electricity hit my brain, I skipped uncontrollably off with the fairies. And it wasn't a nice feeling. In fact, it felt exactly like the zoned-out state I've been fighting all these years and have only recently started to feel more in control of.

I had wondered if this might happen. From what I've read so far about the bits of the brain involved in thinking creatively, it seems that the circuits required are more or less the same as those needed to sustain focus. To think creatively, though, the balance of activity has to be a little bit different. Rather than keeping the frontal bits of the brain in control of attention, to keep thought processes on the straight and narrow, thinking creatively requires a softer focus, a more mind-wandering state of mind.

To help get me into this particular zone, Lila has attached one electrode to my prefrontal cortex, where the current will interfere with its normal functioning, thereby freeing my mind from its sensible shackles for the next 20 minutes. Given that I have spent a large part of the past few months trying to train my prefrontal cortex to be more in control of this kind of thing, it might seem strange to try and nudge it back in the other direction, but there is method in my madness. Increasingly, I have decided that what I really need to do for my brain isn't to pick a circuit and train it to excel in a particular direction — to be honest, that is starting to sound a bit 'last season'. Instead, what I am aiming for is control over the various mental states it is possible to be in. When I want to focus, I can slip into that state of mind. And when I want to think creatively, I want to be able to set it free. I'm hoping that by having my brain stimulated to help me practise this state, I will be better able to slip into it at will — as opposed to wandering into it against my better judgement.

It's an exciting prospect, if I can make it happen. And why not? With Mike and Joe's help, it took just a week to learn to recognise what it felt like to be in the right zone for focusing for longish periods. This lesson has remained with me ever since, making it much easier to knuckle down when I need to. This isn't because the training 'changed my brain' in any physical way that you could point at on a brain scan but because I know how it is supposed to feel when I am working well — that relaxed but engaged state where everything feels just the right amount of easy and the right amount challenging. More importantly, I also learned to recognise how it feels when I am not focused, and have found new ways to get back to it (a long walk, spot of meditation, or a singalong to loud music in the kitchen).

If I can build on this control, by learning how to turn it on and off, my mornings could consist of freewheeling creative time followed by a quick walk to reset, and then an afternoon of focused work to hone my ideas into something useable. This kind of mind control would be an incredibly useful skill for a journalist — in fact, I can't imagine any job where that sort of mental control wouldn't be useful. And my son's fancy dress costumes would be magnificent.

When it comes to creativity, it seems, the prefrontal cortex (PFC) is a kind of filter or gatekeeper that must be tamed by just the right amount. Too much dampening would be less than ideal for doing most things that we have to do. An important part of the PFC's job is to sift though the possible courses of action and select the most appropriate. This is why neuroscientists John Kounios and Mark Beeman, authors of *The Eureka Factor: aha moments, creative insight, and the brain,* call the PFC 'the box', and argue that it is absolutely critical to living a normal life. The prefrontal cortex is the reason why, for example, if you need to jot down a phone number and there is a pen and a lipstick on the table in front of you, you don't need to spend much mental energy pondering which one would be best for the job. Without it, we would have to think carefully about absolutely everything we do, as if we were seeing it for the first time. It would be exhausting and we'd never get anything done. So the PFC is very useful, unless, as Kounios and Beeman point out, you are really wanting to think *outside* the box.

If you want evidence for the benefits of less prefrontal control on creativity, you don't have to look further than the nearest young child. The frontal networks are the last part of the brain to develop, which is probably why kids are so naturally creative. A five-year-old can look at a cardboard box and imagine it being a robot, a car, a rocket, or a boat — sometimes all of them in the same play session. An adult, looking at the same box, might get stuck at 'empty container that is making the house look untidy'.

My son (now six) also specialises in random trains of thought that often make no sense, but that are undoubtedly very creative. A recent example: 'Mummy, did you know that people can freeze? And if they do, they can't go down a normal slide, they have to go sledging. On a block of ice. But if they went over a bump they'd go *wheeee* up in the air, but then they'd hit the corner of the ice cube and that would hurt their bottom ...' An adult with a fully functioning prefrontal cortex would find that it kicked in on 'did you know people can freeze' and shut down the whole crazy train of thought before it got going.

One idea is that this is an evolutionary thing: when kids are young and everything is worth learning, the brain's priority is to be open to all possibilities, however random and unproductive. They are also busy learning language, and so need to make connections between distantly related words and meanings — so the wider the focus the better. Adults, on the other hand, need to make sense of the world quickly and efficiently, because they are the ones in charge, having to make all the important life-and-death decisions. Poor, responsible, boring old us.

Some people, though, do manage to cultivate the best of both worlds. I once interviewed Oscar-winning director and creator of *Wallace and Gromit* Nick Park for a children's radio show, and while he described how he came up with Wallace's weird and wacky inventions, it really struck me that he seemed to think creatively in the same way as a child. People like him who retain this ability into adulthood might have naturally lower PFC activity, or it could be that they have a better ability to switch into and out of the creative mode. My money would be on the latter. Clearly, Nick Park has enough frontal control to turn his crazy ideas into actual films and to run a very successful production company.

So I know what I'm aiming for — control over the switch between creative mode and adult mode. What I don't know is how to develop the kind of mind control that allows you to slip into childlike open-ended thinking and then flip back into sensible mode to decide what is useful-random, and what is just plain silly?

One fly in the ointment: there aren't that many researchers looking at creativity — and those that are still haven't managed to agree on what creative thought processes actually are and how they occur in the brain. A lot of this debate surrounds a disagreement over whether creativity can ever be conscious thinking, or whether it is always a more mysterious, unconscious thing that only shows up when it is good and ready.

On my way to Kansas, I stopped by John Kounios' office at Drexel University, in Philadelphia, for his take on the situation. He told me that, while it is possible to solve problems deliberately and logically, that doesn't, in his view, count as being creative — creativity is always an unconscious thing, an 'aha' moment or sudden insight that pops up, seemingly out of nowhere. They don't really come from nowhere, of course. For old information to connect in new and surprising ways, it has to have been put there in the first place, stored in memory somewhere in the brain. Then, if you're lucky, when you are busy thinking about something else, these distant strands of information will link up unexpectedly and burst through — 'aha'.

The trouble with unconscious thoughts is that you can't just decide that you want to make them happen, Kounios tells me. 'You can't use a conscious strategy to influence an unconscious process. It just doesn't work very well.' I'm reminded of my efforts to re-train my unconscious cognitive bias, and this point does ring true. Consciously knowing that people probably aren't judging me harshly never changed how it *felt*, emotionally, to be in a challenging social situation. Only by working on my unconscious bias using an unconscious strategy was I able to turn it back in a healthier direction. Unconscious practice makes a conscious difference.

Unfortunately, there is no creativity version of happy-face clicking that will get me in touch with my inner genius. What I can do, though, Kounios suggests, is set up an environment where 'aha' moments are more likely to occur. I'll be trying this when I get home to the UK.

Lila Chrysikou thinks about creativity a little differently. She is open to the possibility that creative ideas can come either as a moment of sudden insight after a period of unconscious musing, but equally could come after a period of thinking directly about a problem. 'There's no evidence that one is better than the other for creativity,' she tells me. If this is right, it would be a major bonus to anyone wanting to improve on their creative powers — because conscious, deliberate thinking is something that can be changed if someone teaches you how.

In a recent study, Tony McCaffrey, of the University of Massachusetts, in Amherst, found that after people had spent 20 minutes thinking of familiar objects by listing all of their component parts (for example: 'a candle is made up of a wick, which is string, which is made up of long, interwoven fibrous strands ... It is also made of wax, which is a cylinder of fatty stuff'), they performed better on a test of their creativity afterwards.[1] The idea is that training people to think about not just the obvious features of an object ('a pen makes marks on paper') but about all other aspects of it ('it is long, thin, and hard') helps them to think of ordinary things in new ways. If someone thought about a pen like this and then later found they needed something long and thin to stir some paint, for example, they might consider using the pen. More generally, learning to look past the obvious features of things might become a habit that spills over to other areas of life and work.

Which sounds like it might be workable if you have the time and the inclination to properly break down a problem into its component parts. For his part, Kounios insists that if a skill can be taught then it doesn't count as creativity. But, actually, I'm not sure I agree with him on that — surely the important measure of creativity is what you come up with, not the process by which it happens? I'm not in a position to decide who's right, scientifically. The two camps will have to carry on, as Kounios puts it, 'shouting at each other and throwing experimental results at each other', until there is enough evidence to settle it in one direction or another. All I care about is finding a way to make good ideas more likely to happen in real life, so my plan is to try both approaches: thinking hard about a problem, and changing my environment so that unconscious 'aha' moments are more likely to happen.

As for a test of my creative prowess, I plan to step out of my journalistic comfort zone and try my hand at fiction. About seven years ago, I had a couple of ideas for children's books. Both of these ideas seem okay to me, but, because fiction writing is something I know nothing about, I have never done anything with either. Having another try at it seems as good a measure of my creative powers as any — and, who knows, I might just become the next J.K. Rowling.

First, though, I'm in Kansas, having a go at what could turn out to be the easy option, if electrical brain stimulation (tDCS, or transcranial direct current stimulation) ever makes it safely out of the lab and into our homes. Actually, there is a (literally) buzzing online community of home tDCS-ers, who make their own devices — using a couple of wet sponges and a battery — so it is theoretically possible that I could do this to myself. I don't recommend that anyone try this at home any time soon, though. I can't think of a way to put it any better than neuroscientist Micah Allen, of University College London, who said on Twitter recently: 'I don't know a single neuroscientist that doesn't think it's a terrible idea to strap a battery to your head outside of a laboratory.' How would you know where the current was going? How would you know you weren't zapping the wrong bit too hard, risking a seizure? What if you leave the current on for too long and cook your brain?

It doesn't matter how much anxiety-reducing meditation I do, I will always feel uncomfortable about the prospect of strapping a battery to my head. Today is no exception, even though I am in Lila Chrysikou's lab at the University of Kansas, in Lawrence, and she knows exactly what she is doing.

Lawrence, Kansas, is an interesting place. On one level, it seems like a very strange place to have a university — it's classic, small-town America: a long, wide main street, surrounded by rows of clapboard houses with rocking chairs on the porch. I'm staying in one of these, with a lovely mid-western lady called Karen who bakes cinnamon bread and granola for breakfast and regales me with stories about her kids and grandkids, who have good American names like Brent and Jackson. She seems like the kind of grandma that doesn't take any nonsense from anyone, so no one ever tries. She's fantastic, I could chat to her all day.

About a ten-minute walk from Karen's B&B, at the top of seemingly the only hill in Kansas, is the university. When you discover that Lawrence is home to 27,000 students, the rest of the town makes more sense, particularly the huge number of bars, restaurants, and vintage clothing shops along the main street. I'd love to explore, but I have an appointment to meet Lila, in yet another windowless room — to be tested and zapped for the rest of the morning.

Luckily, Lila is great company, and like so many of the scientists I've met so far, she is apologetic about putting me through hour upon hour of lab tests. Psychology researchers are more used to experimenting on students who need to do some research for course credits and are often tired, hungover, or less than enthusiastic about the research itself. They're not used to visiting enthusiasts wanting to do all their experiments in the space of two or three days. I'm genuinely enjoying it, though, not that any of them seem to believe me.

The main test I'm going to do today is a measure of creativity called the 'Uncommon Uses Task'. Like so many psychology tests, this takes the form of a series of photographs on a computer screen. This time, the pictures are of various kinds of objects: a saxophone, a ski, a wheelbarrow, a shoe, and so on. My job is to come up with a use for that object that is anything but what you normally do with it. I will get marked on how many times I draw a blank, how quickly (in milliseconds) I come up with an idea, and how unusual it is.

In previous experiments, Lila and her colleagues found that inhibiting the prefrontal cortex not only increases the number of ideas people came up with overall, but also made them answer significantly more quickly. Also, the uses people came up with under tDCS were further removed from the standard uses. Yesterday, I completed the task without any stimulation, for a baseline measure, and over the next three days I am to have three sessions of 20 minutes: one a day, and one of which will be the sham condition, where they will turn off the current after only a few seconds without telling me. This is to see whether just *thinking* that I'm being stimulated is enough to make me a creative genius.

I don't know for sure, but I'm guessing that today's stimulation is not the sham, because the best way I can describe the way my head is feeling is: a bit lopsided. I am trying to focus on the instructions on the screen, but my attention keeps skipping away to some point in the middle distance. I'm just hoping I can stick to the task at hand. There are benefits to this feeling, though. I am aware that some of my suggestions are weird, verging on bonkers, and I have to say them out loud so that Lila can measure my reaction time. But I definitely feel less embarrassed about my ideas than yesterday, when I did the baseline measure without any stimulation at all. This could be down to my being more familiar with the task and with Lila, but it might also be that my impulse control, another job of the PFC, has been dialed down a little as well. It's a bit like the effect of a glass of wine, or of being too tired to be tactful. Both alcohol and tiredness dampen down PFC activity — which raises the intriguing possibility that a lunchtime tipple might work pretty well for stimulating an afternoon of creative thought.

In the lab, with my focus turned down and my inhibitions off with the fairies, it feels like a licence to be random, and it's strangely liberating. When a picture of some fishing waders comes on the screen, I suggest that they could be 'trousers for a horse's front legs'. When a kite pops up, I say: 'to carry apples if you have lots'. A flip-flop? A letter rack. A fanblade? A rotatable snack-dish. And so on, and so on. I'm aware that what I'm coming up with is a bit odd, but Lila assures me afterwards that they are all perfectly reasonable. They don't count suggestions that really don't make sense, she assures me, but mine are all just about do-able. You *could* use a flip flop as a letter rack if you really wanted to.

Lila is nothing if not efficient, and processes my results in half an hour while I slip next door to get on with the less-creative side to my job — frantically typing out interviews word for word so that I can quote people accurately later. And to accurately quote Lila when she showed me my results ... 'Ta-dah!'

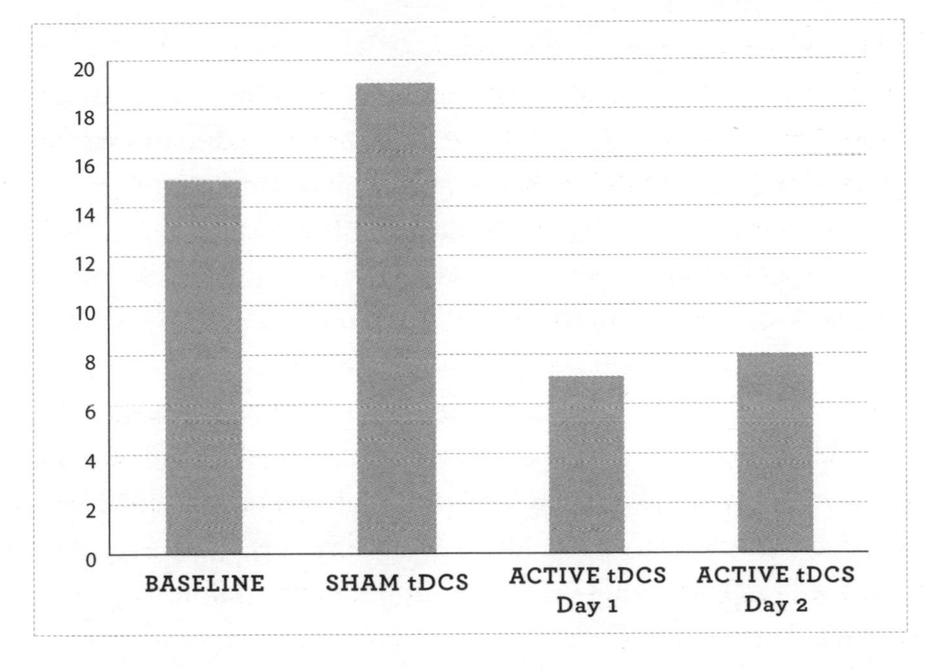


Fig 12. Number of omissions (blank answers).

The number of omissions — where I looked at a picture and couldn't think of a single other use for it in the nine allotted seconds — was reduced by half when the tDCS current was flowing. I'm not sure what sort of effect I expected, but this was pretty clear: having a 20-minute zap in the head made me come up with more ideas. I was also faster — up to half a second faster — to come up with the ideas after I'd been plugged in.

Half a second doesn't sound like that much of a change, but Lila assures me that it's actually quite a big jump. 'To a psychologist, when we are thinking of a difference in reaction time, we are thinking of 50 milliseconds, 60 milliseconds; 500 milliseconds doesn't sound like much, but to a psychologist that's huge.'

Lila is beaming when she shows me this graph, even though she has seen something similar many times in her real studies. 'I'm always super excited when it works,' she told me. 'It's just so weird when it does — but it does!'

I'm amazed, too. Being in the zoned-out state, whether I like feeling that way or not, does seem to be more useful for coming up with original ideas. Maybe my natural state, with few powers of focus, is something to be embraced rather than fought. I'm a little worried that I might have to choose between the two — a state of intense focus that isn't terribly creative, or creativity with no powers of self-control. I'm really hoping not.

Lila later analyses the content of my actual ideas to see whether they got any more creative. This is measured on a sliding scale between uses that are close to the normal use of an object and things that would work but have nothing to do with what the object was designed for. For example, if it was a butter knife and the suggested use was for spreading icing on a cake — for that, you'd get 1 point. If the use was more about the properties of the object rather than what it does — for example, if you suggested using the knife to focus light onto some kindling to start a fire — then you'd get 4 points. I only got one 4, while under tDCS: for suggesting that barbed wire could be used to make jewellery. Overall, though, my score edged slightly upwards while under tDCS, from an average of 2.5 to closer to 3 points. So slightly more creative perhaps, and roughly equivalent with what Lila had found in her previous studies.

I now also have a new term for the zoned-out feeling that I have always struggled to put into words. About the closest I have come up with is being mentally offline or feeling like my engines are revving while I'm still in neutral. Now Lila has given me the proper scientific name for it: hypofrontality — low activity in my prefrontal cortex. And now I know that the days I feel 'a bit hypofrontal' might be a good time to look for ideas, however random they might seem.

Given that this zoned-out state of mind seems to come fairly naturally to me, what I want to know now is how to get into the hypofrontal creative zone without slipping into mind-wandering inertia, boredom, or stressing out about having nothing to show for my time.

This is something that has been a challenge ever since I started out as a science journalist, 15 years ago. In the early days, I tried nominating one day every week as an 'ideas day'. This meant giving myself a whole day off from writing, panicking about deadlines, and trying to get paid, to read around and look for original new story ideas. It sounds like an idyllic way to spend the day, lounging around in some floaty writer's wear, thinking big thoughts while reclining on the sofa. In reality, it is by far the hardest part of the job. A freelancer's challenge is to find ideas that none of the in-house editors have thought of, and that means ignoring the low-hanging fruit like press-releases, and anything in big, famous journals like *Science* and *Nature*, and instead going to strange-sounding conferences, reading obscure scientific journals, and hopping randomly around the web in the hope that a new angle will present itself.

Sometimes this approach paid off. One time, while perusing the website of a marine biology group, I stumbled across an area of research I didn't know existed: scientists recording the burp-like sounds that fish make to communicate with each other. Apparently, you can use slight differences in the sound of these burps, grunts, and whistles to tell haddock from pollock, and cod from herring, which is useful if you want to estimate the populations of each. More to the point, who knew that fish talked to each other at all? That was a good day in the office, and it ended up being a fun piece for a BBC Radio science programme. The researcher even provided impressions of different fish for comic effect.

Most of the time, though, this scattergun approach wasn't so fruitful. It didn't take long to get bored of aimlessly trawling the internet for something that sparked an idea. And after digging my way through pile after pile of turgid scientific papers looking for a gem that no one else had spotted, I'd start to lose the will to live. If you have an attention span like mine, I decided, it's impossible to order creative ideas on demand.

Maybe, though, the knowledge of what hypofrontality feels like and how it helps creative thinking is the way to gain control over idea generation. This could be as simple as choosing whether to bother looking for new ideas or not, depending on what state of mind showed up that morning. Or, more usefully, it could mean practising getting into that state deliberately.

In *The Eureka Factor* Kounios and Beeman offer various tips that, although aimed at creating the right conditions for 'aha' moments to pop up, are also suitable for putting the brain into a hypofrontal state. 'Aha' moments, too, only seem to come when the PFC is snoozing on the job. So, given that I don't plan to make a tDCS kit to use at home, for my next bit of me-search, I'm going to try a few of Kounios' and Beeman's strategies. If they work, they would be far easier to implement than DIY brain stimulation and, at the moment, much safer.

The first suggestion from them is, boringly, to do some work. Ideas or solutions to problems might seem to pop up out of nowhere, but in reality they are always the result of making connections between things or ideas that we just haven't put together before. This can only really work if we give the unconscious something to dig about in. Kounios and Beeman suggest working on a problem using all your powers of analysis and sensible thinking until you get stuck. Then, and only then, take a break. During the break, they suggest, it might be an idea to indulge in some sensory deprivation.

They suggest that because, in experiments that recorded the brain's electrical activity as people solved brainteasers, Kounios and Beeman found that one second before the solution popped into their heads, the visual cortex at the back of the head, which processes visual information, briefly switched itself to a kind of 'offline' state.[2] The explanation for this, known as an 'alpha blink', is that the visual cortex shuts off incoming information for just long enough to allow the solution to a problem to pop through. It's an automatic, neural version of scrunching up our eyes when we're trying to remember a fact — it takes visual information out of the equation and lets the rest of the brain take that share of the available thinking power.

The alpha blink is almost certainly beyond our conscious control, Beeman and Kounios point out — but another way to take your visual cortex offline is to close your eyes or sit in the dark while you are waiting for genius to strike. In fact, Kounios suggests, any kind of sensory deprivation might do the trick: ear plugs, standing in the shower, staring at your feet on a long walk, anything that focuses your attention away from the outside world and directs it inside your head.

It sounds doable, so here's the plan: I am going to dig out the pile of notes, drawings, and scribbles — the children's book ideas that I have written down and stuffed into drawers over the past few years. I will think, sensibly and analytically, about what kind of books I want them to be (picture books for younger kids or chapter books for eight-year-olds), think about the best way to fill in the plot holes that have been bugging me, and think about how I might get someone in the know to look at it and tell me if it has promise (or whether it's frankly embarrassing and I should stick to the day job). And then, when I can't think about it anymore, I'm going to go for a long dog-walk with my hood up, and hope the stories write themselves.

While I'm doing the sensible-thinking bit, I try another tip from Beeman and Kounios: to deliberately think about distant places. Experiments have shown that people told to think about far away places, or the far future, are more likely to come up with new ideas, supposedly because anything that broadens our mind has the side-effect of making us think more broadly in general. With this in mind, I gather various souvenirs from far-flung places and use them as paperweights on the crumpled mass of notes. I add some Korean wooden dolls, and one from Japan, a wall-hanging from China, and some Japanese letter stamps. Maybe, just maybe, all of these oriental influences will make the hero of one of the stories — Jiko, a trainee ninja, who is noisy and clumsy and terrible at hiding — spring into life in my imagination. The other story is about a dog, so just going for the walk might help with that one.

The final bit of advice on getting into an ideas-generating state of mind is to cultivate a good mood. That's not a problem this morning — playing at being a children's author is a fun way to spend a few hours, and doesn't feel at all like work. That's a bonus because, personally, when I'm in a bad mood, there's no amount of comedy in the world that will lift me out of it — it's more likely to make me even more irritated. Nevertheless, in studies dating back to the 1980s, volunteers' moods have been manipulated by either thinking about happy or sad memories, or by watching scary or funny film clips; a good mood made insights much easier to come by in the tests that followed.

This effect, John Kounios tells me, is caused by the way that mood affects a part of the brain called the anterior cingulate cortex, or ACC. The ACC is more active when you are in a good mood, he explained, and is also more active when people solve problems by insight rather than logical analysis. The ACC's job is to monitor the rest of the brain for signs of conflict. 'The conflict could be that I give you a problem to solve and there are different possible paths to solving it. If the anterior cingulate is very active, it can detect all of those different possibilities and it can switch to one of them. And it might be something that is very weak and unconscious and you can still switch to it. That's an insight.' On the other hand, 'If the anterior cingulate is in a low state of activity because you're anxious, then you don't detect all those possibilities, you go with the single strongest one. You do the obvious thing.'

It sounds simple enough: power up the ACC with happy thoughts and it will have enough oomph to root out any daft ideas that might just work. Other studies have found that happy thoughts seem to deactivate the PFC more than sad ones, also suggesting that getting into the creative zone is easier when you are happy.[3]

Good mood alone, though, might not be enough. Other studies have found that a positive mindset has to be teamed with a certain amount of motivation, also known as an 'approach' mindset, where you feel like getting stuck in. Being hypofrontal and relaxed, it seems, doesn't do the same job, because this state of mind doesn't provide enough oomph to put in the necessary mental energy.[4]

Interestingly, the same researchers found that the ultimate enemy of the (creative) state is anxiety. Anxiety, as I have learned already, either scatters attention or focuses it on the wrong things. Which probably explains why my 'ideas days' never worked. The longer I spent trying to focus on whatever scientific journal, researcher's website, or article I was reading, the more stressed I would get about being a terrible journalist with no ideas, and about wasting a day when I should be earning some money.

Put all of this together and it seems that, to stand a chance of being creative, you need to be in a good mood, not too relaxed, and definitely not too worried about whether what you are doing is any good. It's a pretty tall order when you actually have to make a living. And when I read further into the literature on mood and creativity, it gets more complex still. Recent research suggests that whether you happen to be high or low on the neurotransmitter dopamine — a brain chemical that has a hand in everything from attention to desire to switching between one task and another — is important in deciding whether or not a good mood makes you more creative.

According to a recent study by Bernhard Hommel, of Leiden University, in the Netherlands, you only really get a creativity boost from a good mood if you happen to also be low on dopamine. People with higher than average dopamine levels, the study suggested, might get the opposite effect — for them, a good mood might be bad for creativity.[5]

This in turn is complicated by the fact that the whole dopamine story has changed a fair bit in recent years. Dopamine has long been thought of as the 'pleasure' neurotransmitter, which spikes when we get a reward (be that praise, sex, or a hit from a drug). Now, though, researchers are starting to think of it as a chemical that drives 'wanting' rather than being about the pleasure of 'getting'. Which explains why wanting something — cake, wine, whatever — often elicits a more powerful feeling than the pleasure of having it. It's why smokers carry on smoking even though they have started to hate the habit and may have lost pleasure in the taste. Perhaps in the context of creativity, dopamine might stimulate a kind of wanting, searching mindset that fosters creativity.

I'm curious about my own dopamine levels, but with all the overseas travel, I no longer have the budget for another expensive lab test. Amazingly, though, I stumble across a quick and easy way to get a measure of your own personal dopamine levels — with nothing more scientific than a video recording of your face. For some reason, eye-blink rate correlates nicely with actual dopamine levels in the brain: a low blinking-rate signifies low dopamine, and a higher blinking-rate represents higher dopamine. All you have to do is film your face for, say, six minutes, while sitting alone (not talking or reading, as these change your blink rate), just staring at a blank wall a metre away from you. Afterwards, watch the video (if you can bear it) and count the total number of blinks. Divide that number by six and you have a measure of your blink rate per minute. One tip: don't measure your eye-blink rate in the evening — everyone blinks more than usual at the end of the day.

Eye-blink rates vary a lot from person to person, but according to a couple of studies of healthy people, a rough average is 15 blinks per minute.[6] More than 20 blinks per minute is considered high, and below ten, low.[7] I came up with a blink rate of 11: slightly below average, but not mind-bogglingly low.

Incidentally, finding out that my dopamine levels are lower than average sent me off on a tangent about what that might mean for my brain. Only a few clicks away was a website that claimed that low dopamine is linked to: difficulty with sustaining focus; anxiety; impulsivity; and a lack of interest in food (the latter is also true of me; I would be first in the queue if they ever brought out a food-replacement pill). Which made me wonder if dopamine might be underlying everything about my brain that was bugging me. Might that be a better, and easier, target than all this mucking about with mental states?

Unfortunately, attempting to change your own dopamine levels is not straightforward, even if I were to try to only tweak it with creativity in mind. A recent study found that giving people supplements of L-tyrosine, an amino acid that the body converts into dopamine, did boost the kind of creativity that is needed to bring unrelated pieces of information together in new ways, known as convergent thinking. On the other hand, it didn't seem to help people come up with new ideas out of the blue, so-called divergent thinking. In fact, the researchers concluded, for blue-sky creativity, a naturally low dopamine level might actually be better.[8]

So, I'm not sure messing with dopamine levels by taking

supplements is the answer. Learning to use hypofrontality flexibly but without losing control of your focus sounds like a far smarter use of what nature gave me.

Another option is to take Kounios' and Beeman's advice and use the parts of the day when you are naturally less alert, for thinking creatively — evenings for morning people, mornings for night owls; or, if all else fails, sitting in a dark corner, closing your eyes, or wearing earplugs will most definitely set the mind wandering. With no external stimuli, the restless mind can't help itself. You might come across a bit weird in the office, but think of it as part of your new image as a creative genius.

As it happened, I never did get as far as the mind-wandering dog walk. Once I dedicated the morning to it, and started thinking hard, the ideas started flowing, complete with a few aha moments: 'Jiko has ADHD!' This then sent me off down other avenues, like how a boy with ADHD would be a terrible ninja on the one hand because he couldn't sit still, but on the other he would be great at anything involving jumping about and running around at top speed. *Aha!* The moral of the story is: play to your strengths!

I bashed out a 500-word story, sent it to my agent and waited (this story, and another I tried out, are at the end of this chapter). And ... no luck yet — watch this space. But as Lila gently pointed out to me when we first met and I told her about my children's book plan, 'You can be more creative, but it has to be good, right?'

## THE PREFRONTAL CORTEX: GATEKEEPER OF GOOD IDEAS

This is where the story takes a bit of a turn, because more recent research has shown that, while hypofrontality is the perfect state for generating ideas, the prefrontal cortex is most definitely needed to sort the wheat from the chaff. As my son demonstrated in his reveries about frozen people, it is perfectly possible to be creative in a way that makes no sense to anyone else. Which is fine when you're six and free-forming about nothing in particular, but if you're an adult who wants to apply their newfound creative skills to something useful, hypofrontality alone isn't going to cut it. In fact, a little PFC action can go a long way towards making creative thoughts work in the real world.

Back in Kansas, Lila and I are headed to Kansas City, about an hour's drive from Lawrence, where the university houses its medical centre. I'm booked in for another MRI scan, and, for the first time, will get a snapshot of it in action while I try to apply creativity to solve a problem, and watch my PFC as it (hopefully) dials up and down to allow for a useful level of creativity. To do this, we are using functional MRI, which gives an indirect measure of brain activity by measuring blood flow to parts of the brain that are doing the work. I'm going to try Lila's latest experiment, which is an attempt to measure creative thinking in a more real-world kind of problem than the unusual-uses test. 'In real-world problem solving, rarely are you given objects and asked "tell me what else can you do with it",' Lila tells me later. 'You are given a problem or a goal. So you might want to take your dog for a walk and the leash breaks. You might think, "Oh, I can use my belt." You don't sit around all day thinking of all the things you could do with your belt! So it's the other way around.'

An emerging idea in the creativity research world is that in this kind of problem the PFC has a dual role, and hypofrontality is only the first part of the story.

Until recently, the received wisdom was that the executive control network and default-mode network worked against each other — executive control focusing attention outside of the self while the default mode focuses inwards. It's certainly true that you can't have two focuses at the same time, as far as anyone knows. What the brain can do, though, is flip from one state to another, sometimes very quickly. The experiment I'm about to do is designed to watch that happen in real time.

In the scanner, my job is to think about a particular goal; for example, to make a fire. Then a one-word example of how to achieve the goal flashes up (newspaper), followed by another two (a pen and a pencil). I need to think creatively about which one of the latter two items could best be used in the same way as the first to achieve the set goal. In this case, I would have to choose the pencil, because it would burn better than a pen.

Doing this exercise in the scanner should reveal what parts of the brain are involved in thinking about everyday objects in unusual ways, depending on what it is you are trying to do. It's a simplified version of creative problem solving for the purposes of identifying how the brain goes about the job.

The good news is that my prefrontal cortex is in perfectly decent working order, and lights up in all the right places at the right time. The bad news is that this new understanding about the prefrontal cortex's role in creative thinking is only going to make it more difficult to order ideas on demand. According to Lila, highly creative people are probably able to flick between low and high prefrontal activity very efficiently, sometimes quickly and sometimes slowly. And nobody knows how they do it.

'People who might be able to slip in and out of these states, and also be able to control the speed with which that occurs — maybe not intentionally to order, but it might be a natural process — these are the individuals who might be very successful.'

This theory is all very new, and nobody has any bright ideas yet about how to practise this kind of control. The best we've got so far, research-wise, are the studies of expert meditators, who have seemingly inhuman powers of frontal control — and no one knows if that will help specifically for creativity. My meditation practice has dropped off a little recently, but this does make me think hard about adding it back into the daily routine.

Hypofrontal creativity is probably easier to achieve than rapid switching between low and high levels of mental control, because everyone can easily recognise the feeling of being zoned out and off with the fairies. Recognising that state and letting the mind wander where it will could be the best chance so far of coming up with solutions to problems or with ideas that the world doesn't yet know that it needs. If you're lucky, your assessment of the ideas when you're more in the room, with the PFC back in charge, will help you sort the good stuff from the junk.

It's not exactly the answer I was hoping for, but as yet we don't know enough about how switching works to find ways to make it happen. I ask Lila if she thinks it will be possible any time soon. 'We know pieces of the puzzle, but we haven't put it all together. We haven't examined what happens when you increase or decrease your control and whether you can do it on the same person for the same task. That's ultimately the challenge.'

And what about brain stimulation to get into the hypofrontal state? She looks pained, and tells me about a headline once written about her work that screamed: *Prototype headband makes you more creative!!* We laugh at how ridiculous it is ... then I ask anyway if it might just work. 'It's not that we don't want to apply it, it's just that we don't want to suggest that this is ready for prime time,' she says. 'I wouldn't sign and say "Yeah, everybody that uses this, this is gonna happen." No.'

So far, this much we know: brain stimulation that gets people into a hypofrontal mode helps with the generation of new ideas. We also know from research into meditation, and the kind of cognitive training I did in Boston, that it is possible to improve control over attention and focus. What we don't know is whether practising both of these separately will help anyone switch between the two.

'There is no one answer for everything,' says Lila, with a shrug. 'For better or for worse, we are very complex beings.'

One thing that does occur to me, though, is that we, the wide-eyed users of all this neuroscience research, have been barking up the wrong tree with the whole 'use it or lose it' view of neuroscience. Training a particular bit of circuitry — even if such a thing is possible — might not be what we want, after all. Instead, the picture that is emerging is that, to build brainpower, it's not muscle you want, it's flexibility.

## MY KIDS' BOOK IDEAS ...

### *Dave's Doggy Detective Agency, and the Mystery of the Missing Pups*

Molly is a six-year-old dog who has learnt to speak 'human' by listening to her people and to Radio 4 when her humans go out. She never uses it with her humans at home because people like dogs to listen and lick their hands, not talk back.

When her main human gets a job, they hire a dog walker. Unbeknown to them, they hire Dave, who runs an undercover doggy detective agency masquerading as a dog-walking service.

The mystery: dogs have been going missing from the local park and outside shops. People are really upset: the adults talk in hushed voices about dog-fighting rings and the children panic that their beloved dogs have been made into fur coats, *101 Dalmations*-style.

Dave puts Molly — his best doggy detective — on the case.

The dogs have been stolen by a lazy person, whose doctor has given her a treadmill to power her TV. It's the only way the TV will work now. The doctor did it as a last resort after trying to persuade the lazy person to not just watch TV all day. However, instead of exercising, she sent her son out to steal dogs instead. She keeps the dogs in cages in the basement, bringing them up only when it's their turn to run on the treadmill. She has more than one dog, because 'otherwise it'd be cruel'. Her son is kind, feeding and looking after the dogs, but is too scared of his mother to disobey her. If he did, she'd make HIM run on the treadmill all the time.

Molly and Dave track the dog thieves down, and help the boy ... aided by Billy, Molly's youngest human; the other doggy detectives in the team; and Governor, the black cat who is the team's eyes and ears at night.

### *Jiko, the Noisy Ninja*

In the house with the dirty windows, everyone was sleeping.

But then ...

BANG!!!

CRASH!!!

'Uh oh,' said Jiko. And he jumped into the shadows ...

Jiko (Japanese for 'trouble') is a boy who has just finished ninja school, and is training to be a house ninja for Sam's family. It's a little-known secret that all homes with children in them have a ninja who secretly helps around the house. Grandmas employ them to help out because they know how messy children can be. Mums don't know about them, though. They wouldn't like it, because mums like to Do It All.

Jiko, though, has ADHD, and although he is brilliant at moving fast and doing things with enthusiasm, he is clumsy, easily distracted, and not very good at sitting still for long enough to hide.

His teacher, Mr Soso (Japanese for 'neat'), despairs of him and wonders if he will ever slow down enough to become the great ninja he knows he can be.

In this story, Jiko makes a big noise at night and accidentally wakes up Sam, the boy in the house. They become best friends and get up to all kinds of mischief together. But will they be able to keep their adventures a secret from Mum, Mr Soso, and the ever-observant Grandma?

# PART TWO
# SPACE AND TIME

CHAPTER 4

# Lost in Space

*'Not all those who wander are lost'*

— J.R.R. TOLKIEN

I'm wandering around Berlin in the rain, with only a soggy map to guide me. There's no doubt about it, I am definitely lost. It doesn't matter how many times I mentally retrace my steps or turn the map upside down, I still have no idea whether to go left or right at this junction. The street sign in front of me is pointing completely the opposite way to what I was expecting, and it has put my sense of direction into a tailspin. Ironically, I'm on my way to meet a navigation researcher, who has brought me a gadget that she says could drastically improve my sense of direction in just six weeks. To meet her, all I need to do is work out which way to go ...

I have high hopes for this particular challenge. Every brain skill I have worked on so far has been underpinned by the question of *whether* it can be changed by working on a specific bit of circuitry. Thanks to psychologist Eleanor Maguire and more than a decade of her team's work with London's taxi drivers, though, navigation seems to be a done deal. Navigation ability can most definitely be improved with training, and this most definitely increases the size of the bit of the brain involved in the task. Over the past 14 years, Maguire has shown time and again that spending a couple of years obsessively learning the layout of London streets increases the size of the hippocampus — a key brain area for spatial navigation. I will find out later that even this well-worn exemplar of neuroplasticity isn't as straightforward as it might be, but for now, lost in Berlin, my hopes of changing my brain to be more *taxi driver* and less *lost cause* are high.

The only question I had was about how to do this to my own brain. I could apply for a London taxi driver licence and spend the next three years memorising street maps, but let's be honest: I don't have the time. And anyway, when I emailed Eleanor Maguire to see if she would be willing to check my skills out and help me improve them, London-cabbie-style, I got a very polite but pointed reply from her secretary saying that, sorry, but Professor Maguire is way too busy to have me in her lab any time soon. 'How about a phonecall?' I asked. 'It should take less than an hour.' Sorry, came the reply, the diary really is very full for the next year. It was all very polite, but the take-home message was clear: Not. Interested.

Cue a frantic search for other researchers doing work in spatial navigation — and, in hindsight, I'm very glad I did, because before long I found something that sounded even more intriguing. A group of researchers in Germany is working on a very different approach to the problem of navigation — concentrating not on improving on what nature gave us, but on *adding* an extra sense, to test whether the human brain can assimilate it as one of its own. In this case, they are tying to add the ability to navigate using the Earth's magnetic field. Pigeons can do it, migrating turtles can do it, and if these people have their way, it'll be an optional extra for anyone, like me, wanting to improve their sense of direction by giving the brain a new bit of information that it can't access naturally.

Quite apart from the fact that this sounds incredibly cool, a bolt-on sense of direction sounds like something I could use. I spend a lot of time in the outdoors trying to wear out my young sheepdog so that he'll go to sleep and let me get on with some work, but my sense of direction is so bad that I am too chicken to take him anywhere new on my own. On weekends, we go all over the place. My husband, Jon, has a freakishly good sense of direction, always knows where we left the car when we are out walking, and has an uncanny ability to take shortcuts to get back on the right path. I, on the other hand, take a few turns out of the car park and it feels as if I've been blindfolded and spun around on the spot. Ask me to point back to the car and you might as well ask me to fly to it. So although I have the benefit of working at home and a good chunk of the English countryside to explore every day (and now, the need to reset my attentional focus and/or think creatively as the perfect excuse to get out in it), I don't, because I am too scared of getting lost in the hills with no phone reception, and having to consider eating the dog.

Which is why, on a rainy September day in 2015, I am meeting Susan Wache, part of the team responsible for the feelSpace belt — a prototype navigational aid that constantly computes the direction of north and translates it into a vibration on the northwards-facing part of the wearer's waist. I find her eventually, by applying my usual strategy in these situations: pick a direction, start walking, and hope for the best. When I finally got to the corner where we agreed to meet, I was so relieved that I hugged her like a long-lost friend. I think she was a bit surprised.

The feelSpace is an expensive bit of kit, and it has taken months of badgering via email to persuade Wache's boss, Professor Peter König, of Osnabrück University, to let me take it home for the next six weeks. According to their studies so far, six weeks should be long enough for my brain to adapt, if indeed it is going to. I have Susan to thank for König's change of heart. She is, I soon discover, a card-carrying non-worrier and optimist. When we meet, I have just finished the Oxford study, and we get chatting about everything I have been doing to change my negative mindset. She tells me that she never thinks the worst and she never worries. 'Why would I think I'm a terrible person?' she says, looking like it's the craziest thing she's ever heard. 'I'm not a terrible person!' Thanks to the force of her charm, enthusiasm, and willingness to trust a complete stranger with 2000 euros' worth of kit, König finally agreed — so here I am for a crash course in how to use it.

Susan runs me through the features of the belt over a couple of enormous ice-cream sundaes in a mall in central Berlin. Apparently, afternoon ice cream is very much a thing here — so it sort of counts as a cultural experience. It's not quite what we had planned, though. Having spent the morning trying to navigate the sights with only my brain and a cheap map from the tourist office, this afternoon I'm supposed to be continuing my tour but with an entirely new sense — the ability to feel the pull of magnetic north. Torrential rain, though, is not great for battery-powered bolt-on senses.

So we order the least enormous-looking sundaes and settle down for a chat about the belt. She tells me that, in their previous experiments, after wearing the belt for six weeks, volunteers had a vastly improved sense of direction, and had been able to rewrite their own internal maps, making everything align to magnetic north. Susan, a student in the lab at the time of the study, was one of the research subjects and by the time we meet has been wearing the belt on and off for about six months. She says the effects far exceeded her expectations — and even now she takes it with her whenever she can. 'I was expecting to get a better map in my head from an aerial point of view, but that's not what happened at all. I got a better map, but it is like I am in the map. Like Google Street View but everything is transparent,' Susan told me. It all sounds fantastically sci-fi.

So we chat and eat, and eat and chat, and the sundaes never seem to get any smaller. Finally, when we are both utterly stuffed and feeling a bit sick, Susan whips the belt out of her rucksack and plonks it on the table. She tells me that they are working on a sleeker version for the consumer market — which should be launched fairly soon — but what I'm going to be trying is one of the research prototypes that they use in their studies. Immediately, I see why one of the volunteers had trouble getting it through US customs. It couldn't look more like a bomb belt if it tried. It's a thick, black webbing affair; about three inches wide; with several bulges the size of cigarette packets around the hips, and two ends of a wire hanging down; topped off with an ominous-looking connector. To turn it on, you have to strap it around your waist, zip it together, and connect the ends of the wires — presumably with your best innocent face on.

Some of the bulges are haptic motors, Susan explains, which create a similar kind of buzz to when your phone is on vibrate. One of the bigger rectangular bulges is a GPS unit, which is wired into the haptic buzzers. The GPS constantly monitors the position of north, sending that information to the buzzers, so that the one pointing north vibrates. The other two large bulges on the belt are the batteries, she tells me, and hands over an industrial-looking charger that I'll need to take with me.

Maybe I have spent too much time on the paranoid streets of London, but as Susan stands up, attaches the belt and, totally unselfconsciously, connects the two ends of the wires, I think that the people of Berlin are a bit too laid back for their own good. I glance around us nervously, as the belt buzzes into life, but no one else bats an eyelid. Okay, so Susan is young, in her 20s, with bright ginger hair and an innocent face that matches her outdoorsy outfit, but if I saw a stranger doing this in a very public place, I'd run for the hills.

Then it's my turn. I hold my clothes out of the way as Susan zips me into the belt and connects the wires. I'm less than relaxed, and remark that I definitely wouldn't do this in the middle of New York or London. She laughs. 'This is Berlin, nobody cares.' Really? I'm sure I saw an armed policeman earlier ... Back at my friends Neil and Jess' apartment, where I am staying, and wondering if I'm just being weird, Jess (who is American) agrees with me. 'You'd get tasered wearing that in New York,' she assures me. 'At best ...' I add, darkly. I've already decided not to take it to Chicago with me next month when I go to a big neuroscience conference. Much as I'd love to try it out in another unfamiliar city, it's probably not worth the risk of being shot.

Since we don't have the option of exploring Berlin's historical sights this afternoon, Susan and I head instead to the Sony building — a massive covered square with a cinema on one side, where she tells me they hold the big movie premieres in Berlin. It's large enough to do the kind of navigation tasks that they do in lab studies to test people's sense of direction, and we can try them with and without the belt. These are a bit like those trust exercises you do on corporate bonding days. First, we turn the belt off, I close my eyes and hold Susan's arm as she guides me around, taking lots of turns. After a minute or so, we stop and she asks me to point to the cinema. I do okay, pointing more or less in the direction where it is. There's a big fountain in the middle of the square, which makes enough noise to help me vaguely keep track of where I am, which probably counts as cheating.

Then I turn the belt on, make a mental note of where on my

body the belt is buzzing in relation to the direction of the cinema, and then close my eyes. Again, Susan walks me around in circles, making lots of turns and double-backs, and then stops and asks me again to point to the cinema. This time it's so much easier. Before I was guessing, this time I know I'm right. 'It's over there,' I say confidently, and I'm right. It's almost too easy.

With that we say our goodbyes, and I take my new sense of direction back to Jess and Neil's, apartment, which I am now very much aware is north-east of the Sony Centre.

The belt is big, heavy, and I feel incredibly self-conscious, but I quickly start to enjoy having a secret sense all to myself. That evening, though, when Neil and Jess show me around their neighbourhood, me with belt attached, I discover that I am not the only one in the group with a mental compass. At one point, Neil gestures down a street and says, 'We need to go this way.' 'Aha!' I pipe up, feeling very pleased with myself. 'You mean we need to go north?' 'Yes,' he says, raising an eyebrow at me, as if that part was obvious. I had never considered that people might learn the position of north and use it to navigate in familiar environments. It's just not something I would naturally think of doing.

The more I start talking to people about how they navigate, though, the more it becomes clear that what seemed to me like a freakish knowledge of where north is, held only by a few people I know, turns out to be a fairly common strategy. Unlike turtles, pigeons, and, according to a handful of studies, perhaps even cows and dogs, humans don't, as far as we know, have an on-board compass, so knowledge of cardinal directions (the points of the compass) is only ever a learned skill.

My husband, Jon, demonstrated how he uses this method when we stopped for a quick game of 'guess which way is north' while we were out on a dog walk. 'I know that the main road is over there, running that way, and that our house is south of it, and we came this way to get here, so north is over ... there.' It made absolutely no sense to me at the time, but my hope is that with the help of the belt, I'll be able to use this kind of thinking, too. Once I've learned the general lay of the land in relation to points of the compass, I should be able to mentally map everything in the environment in relation to that.

I want to know more about human navigation skills, so I delve into the murky world of scientific literature and find something that I hadn't been expecting: what I have isn't, strictly speaking, a lack of navigation skills — I use one of two bona fide navigation strategies. Some people, like Jon and Neil, navigate by calculating the geometric layout of an area and linking them together to make a mental map. According to the research, this strategy is more commonly used by men, and is useful if you need to keep a general gist of the landscape in mind. If you have a mental map in your head, it makes shortcuts much easier, as you can make an educated guess about which way is home. It also means that if you find out where north is, you can add that to your internal map, too.

Others, like me — and if research is to be believed, women in general — are much more likely to remember specific routes based on landmarks: follow this road until you reach the church, cross over and turn right at the next traffic lights — that kind of thing. The problem with this strategy is that it is harder to take a short cut, because as soon as you are off a known route, you're lost. It may explain why I have a reputation in my household for suddenly exclaiming 'Oh, we're *here!*' when we reach a particular street corner. At which point Jon looks at me, confused, and says, 'Err, yeah. Where did you think we were?'

This 'landmarks' strategy might actually work better in certain circumstances. In experiments, women tend to outperform men in navigating by landmarks and also generally have a better memory about what landmarks look like.[1] And if you have enough landmarked routes banked, then making shortcuts between them is presumably less of a problem. Still, relying on landmarks with no mental overview of where you are and where you are going is far more likely to result in the kind of sense-of-direction failures that I have.

As an aside, one theory for why these sex differences exist has something to do with the way that hormones act on the brain. Differences between the sexes don't appear until puberty, and women get better at spatial navigation in the first week of the menstrual cycle, when oestrogen is low.[2] One idea is that women's brains evolved to specialise in skills that made them successful gatherers: remembering the detail of what good foraging areas look like and where they are in relation to each other. Men's brains are supposedly adapted to be better hunters. They'd need to travel further, perhaps, keeping track of where they are in a wide, open landscape, and know how to make a beeline for home if a predator suddenly appeared, or how to save energy by dragging a woolly mammoth directly home rather than all around the caves.

This theory is clearly untestable, and in this day and age, it actually doesn't matter. The navigation skills I need have nothing to do with finding nuts and berries and everything to do with not getting lost in the countryside.

One thing that does give me hope that I can improve on what I've got is that studies have shown that the best navigators are those who switch from landmark to mental-map strategies, depending on what works best in the circumstances — so, again, it's all about mental flexibility. In theory, then, if I can learn to mentally map my surroundings I'll have a winning combination: a killer memory for landmarks, added on to a detailed cognitive map that is aligned to north. A landmark specialist with a sense of direction might turn out to be a very formidable navigator indeed.

So far, though, I'm just guessing about my navigation strategy, and although studies have shown that people are generally pretty good at assessing their navigational ability (you can test your own using the Santa Barbara Sense of Direction Scale[3]), I'd like to know for sure. And anyway, I'm starting to get very curious about how well my brain's navigation system is working, and that would mean getting a look at my navigation circuits in action while having my brain scanned.

Sadly, as well as Eleanor Maguire not being terribly interested in my hippocampus, her colleagues in the famous and eminent navigation group at University College London have told me that I'm too old to take part in their current brain-scanning study, and they are too busy and too lacking in funds to squeeze me in as a favour. So, while intriguing answers potentially lie a 45-minute train ride from my house, I find myself on another transatlantic flight: this time, to the lab of Russell Epstein, an equally eminent navigation researcher at the University of Pennsylvania, in Philadelphia. When I cornered him at a neuroscience conference in Chicago, he very kindly agreed to scan my brain for clues, and to test my navigation strategy. Once I know what my brain is doing and which bits are dominant, I can start to train it accordingly. At least that's the plan.

I have just two days in the lab in Philadelphia, and Russell and his team have lined up a packed schedule of experiments, some of them in the brain scanner. Luckily, the jetlag is working in my favour, and I'm up at the crack of dawn and raring to go for my first appointment with Steve Marchette, a post-doctoral researcher in Epstein's research team, for the first of several experiments designed to test my abilities.

Steve explains that there are two solutions to the problem of orienting yourself in space. One is to notice where objects are in relation to yourself — the chair is in front of me and to my right, for example. This is an 'egocentric' strategy. The other way to do it is to take note of where things are in relation to each other and the space they are in — the table is about a foot away from the window, and the chair is at the side of the table that is closest to the door. This is called an 'allocentric' strategy.

The first test Steve has for me is to work out which one of these strategies I tend to use. He stands me in front of a table that has a 4 × 4 grid of roughly 20-centimetre squares marked out in yellow tape, and asks me to close my eyes. Then he puts pictures of four objects (for example: a basket, a toy car, an apple, and a typewriter) in some of the squares. I have ten seconds to study the pictures, before I close my eyes again while he takes them away. Then Steve asks me to do one of three things. Either a) put them back where they were, b) move to another side of the table and recreate what I saw in front of me, or c) move to another side of the table and put them back in the same squares that they were in the first time. This last one is the hardest, as I need to mentally rotate the table and remember what went where.

When I think I've got the hang of that, he moves the table out of the way to reveal a larger scale version of the grid on the carpet. This time, I am to do the same task but standing in the middle of the grid rather than on its borders. Sometimes he asks me to recreate the view, other times to put them back in the same squares as before, but with my body facing another direction. Strangely it's even more difficult to do this when I am in the middle of the scene and trying to imagine it from another vantage point — I find myself experiencing the same kind of mental head-spin that I got while I was lost in Berlin. I think I can guess which way this result is going to go.

Then it's off down the corridor to try navigating in a virtual environment. A lot of navigation research is done using adapted video games, and I can't help wondering if that's why I was considered too old for the London study: teaching the average 40-something to move around in a computer game is like pulling teeth compared with your average 20-year-old. Poor Steve. I get it in the end, though, and he explains that I will have several tours of the virtual environment (a fairly featureless maze) to learn where various objects are (a wheelie bin, a chair, a fridge, and so on). Once I've had a chance to learn those, I'll be timed to see how quickly I can get to one of the objects from somewhere else in the maze. I later discover that the point of this is to see whether I can hold a map of the maze in my head so that I can take shortcuts from one object to another, or whether I stick to the routes I know. I can guess the answer to this one, too. I know how to get from the wheelie bin to the chair, because that's the route I learned — but free-forming from there to get to the fridge? Forget it.

Then comes a different kind of virtual-reality game, with a different Steve — post-doctoral researcher Steve Weisberg — who is interested in how and why people's navigational skills vary. This virtual environment looks more like a real place, and apparently is based on a real college campus somewhere in America. This time, I have to learn two routes around the campus and memorise four buildings on each. Then I get to learn two more routes that show me how the first two are connected. My task then is to stand at one building and point to where I think one of the others is. It's easy enough when I'm standing in one route and pointing to a building on the same route, but when I have to point to a building on the other route, which involves mentally computing how the two bits of the campus fit together, it feels like a total shot in the dark.

I'll have to wait until tomorrow afternoon to find out what all this says about my navigation ability, but I'm starting to get the idea that I'm not a great mental-map-maker. I kind of suspected it already — so I'm intrigued to get into the brain scanner this afternoon and find out if there is a physical reason why that might be.

## SOME BRAIN STUFF: THE BRAIN'S GPS

I got the basic rundown of the hippocampus and other important brain areas back in Berlin from Klaus Gramann, of the Berlin Institute of Technology. I went to see him the day after picking up the belt from Susan — and it worked a treat in helping me get there. Gramann's office is in a corner of the city that isn't easy to get to from Neil and Jess' flat, and I needed to take several trains and buses to get there. The night before, I had studied the map and written directions that made use of the belt — things like: 'Leave train station and head east on the number M45 bus. Go ten stops and then walk north on Marchstrasse.' I would never normally consider adding compass directions like this, so I'm intrigued to see if it helps.

As I walk out of the train station, an M45 bus pulls up at the stop next to me. Normally, with ten minutes to go before an appointment and feeling slightly stressed, I'd jump on the first bus that came and hope for the best, banking on having a 50/50 chance of going the right way. This time, though, the belt is buzzing on my back so I can instantly tell that this bus is heading west — the wrong direction. I let it go past, cross the road, and wait patiently for a bus going the other way. I am so pleased, I want to tell the person in front of me in the queue all about it, but my German isn't up to it. I content myself with a self-satisfied grin and the knowledge that I probably just saved myself a very late and flustered arrival.

Ten minutes later, when I get off the bus and find myself at a complicated crossroads, I check my notes again. They tell me to head north, so I turn my body until the buzzing is on my belly, and stride confidently on. It's so easy! I still arrive a few minutes late, but not nearly as late, flustered, and sweaty as I could have been, and so often am. It feels like a whole new world.

Klaus Gramann looks like a scientist as Hollywood might imagine one to be. He's tall, with artfully greying hair raked into a messy quiff, and the kind of laconic smile that reminds me of someone famous that I can't quite put my finger on. I later click that it's Billy Bob Thornton — he's a dead spit, right down to the matching goatee. His office, too, looks way cooler than your average scientist's. A model of a skull, holding a USB cable between its teeth, sits on the desk, next to a bottle of red wine which acts a paperweight on the obligatory pile of scientific papers. In the corner is what looks like a shrink's couch, overlooked by a skeleton wearing a university lab coat and grinning madly as only a skeleton can. He even has scientific gobbledygook written on the blackboard.

In less Hollywood style, though, he seems a little stressed about making the English visitor a decent cup of tea. I assure him that the only tea he can find (loose leaves hand-picked in the Taiwanese mountains) will be absolutely fine, and that I'm sure it will be delicious even without milk, and then we settle down for a rundown of the brain's on-board navigation system.

He tells me that the brain areas involved in navigation converge on the hippocampus, but it is far from the only player. This, like so much about the brain, quickly gets complicated, so he cracks open the plastic skull on his desk and pulls out the plastic brain inside, pointing to each bit as he goes.

In a nutshell, he tells me, the following areas are crucial to finding your way around: first, the parietal cortex, a kind of information hub at the top and back of the brain, integrates incoming information from the eyes, ears, and other senses into a seemingly seamless representation of the environment relative to where you are in it.

The parietal cortex feeds into the hippocampus, more correctly called the hippocampi, because the brain has a matching pair, each of which is a longish looping structure, buried under the wrinkly outer cortex. The hippocampus and the areas around it are home to specialised neurons that construct a mental map. These neurons are the place cells in the hippocampus, and the grid cells and border cells, which live next door in the entorhinal cortex. Grid cells fire in patterns that give a kind of co-ordinates system by which we can keep track of the general layout of our surroundings. Border cells help us compute where one thing ends and something else begins, which is always useful if you want to move between them.

Place cells, though, sound like some kind of computational wizardry that I find difficult to get my head around. They are dotted around the hippocampus, and each is tuned to only fire in a certain location in any given environment. Some of their firing thresholds overlap, so that, as you move through an environment, there is a wave of activity that tells you that you are moving. If you go to a house you've never visited before, a certain place cell will spring into action when you are near the sofa, while another fires when you are by the back door. This mapping happens very quickly and, without us noticing a thing at all, they settle into a firing pattern that allows us to move through the house with no conscious effort. The firing patterns get honed with learning, so the maps get more detailed the more time you spend in a particular place.

The best thing about these place cells, though, is that they are completely re-mappable, so you don't need a new one for every place you ever go. The place cell that tells you where your bed is at home might fire by the photocopier at work and by the cheese aisle in the supermarket. They also fire when you recall a memory of those places. Which sounds like it might get confusing if you were thinking about going to bed while standing at the photocopier, but actually it's a bit more complicated than it sounds. It's the particular firing pattern of a cluster of different place cells that means you can tell different places apart in your mind's eye.

These two bits of information — what your senses are telling you about where things are in relation to you, and what your place and grid cells are telling you about the scene in general — get put together in the retrosplenial cortex (RSC) where another type of neuron, called 'heading cells', compute which way you are facing with relation to the environment. The RSC's main job seems to be to act as a kind of translator that puts you in the picture, so you can move around in an environment efficiently.[4]

Finally, there is the occipital place area, at the back of the brain. This has only recently been discovered, so no one knows exactly what it is doing, but it seems to respond more to places than other things like objects or faces.

When I start to wonder aloud if I might be missing any of these parts, Klaus doesn't seem terribly convinced. 'We all have the same neural basis for all these systems. We have that encoded genetically ... You don't have to do anything, it's just there,' he says. It's a fair point — despite the fact that, thanks to maps, sat nav, and smartphones, pretty much nobody in modern life uses their navigation skills as nature intended, the kit is still in there, waiting to be used. In fact, we use it every day without even realising it. People who have a brain injury or stroke in the brain's navigational centres find it impossible to find their way around in everyday life — sometimes even getting lost in their own home. Some people have this problem from birth for reasons that aren't entirely clear yet. If the system isn't working at all, you'd certainly know about it.

There are definitely differences in how well people use the

various bits of kit, though, and Klaus' research largely revolves around finding out why that might be. If you take a group of people to an unfamiliar environment, for example, they use very different strategies to find their way around. 'Some people create a survey knowledge from their very first encounter. Some never do. So the question is: why is that? If everybody has the same system, why would there be differences?'

Klaus tells me that some of these differences are probably genetic: while we all build brains to more or less the same genetic brain-building programme, brain regions, like noses and bottoms, come in different shapes and sizes. Some people might come better set up to encode the information in the parietal cortex accurately, for example, which gives the rest of the system a good picture to work with. Some people are also better are keeping track of where their body is in space and how quickly it is moving. Some people come with a larger, or more efficient, hippocampus.

Then there are the effects of culture — and it's those that interest me most, because they suggest that hacking the system is at least a possibility, whatever genetics has given you. It's also something that Klaus is interested in. 'If you grow up in the tundra in northern Scandinavia, it's going to be a totally different environment than growing up in New York City,' he points out. In New York, you might be better off with a landmark- or route-based strategy, for example, but in the tundra you'd need to track the sun or plot your position based on the features of the landscape. And if the strategy you develop as a child in a particular culture works, why would you bother developing the other? Over time, these changes add up to different amounts of grey and white matter in the key navigational areas of the brain.

These differences seem to start very early. In experiments that compared the performance of Dutch and African children on the same sort of grid-test that I did in Philadelphia, African children would put the objects back on the grid the way they were with respect to north, south, east, or west, whereas Dutch kids would place them with respect to their own body.

So could an adult urbanite develop the skills of a desert hunter-gatherer? Perhaps. But it has been found that people who stubbornly use a landmark-based route-learning strategy have less grey matter in their hippocampus than people who use a map-based one. There seems to be a trade-off with another part of the brain, the caudate nucleus, which is larger in people who do route-based navigation, and smaller in the map-makers. But the best navigators of all are those who have a similar volume in both the hippocampus and the caudate nucleus, and are able to select the best strategy depending on what the task needs. As with creativity and sustained focus, for navigation skills, building muscle is not the right analogy — again, it all comes down to flexibility.

Klaus suspects that the difference between strategies comes down to differences in activation in the retrosplenial complex — the translator between mental maps and the sense of where the body is in space. This is very much a work in progress, and although I have been hoping to join in with some of the team's latest experiments, it looks like that isn't going to happen. One experiment I wanted to try isn't ready to run, and the other involves walking around a huge hall with a mobile brain activity monitor (EEG) strapped to my back and wearing a virtual-reality headset. It sounds fun, but the hall is being refurbished and won't be useable for at least six months.

A month later, over in Philadelphia, though, I get my chance to look at my retrosplenial complex, and other navigational bits and bobs, for clues. It's off to the brain scanner with me.

The scanner is in another building from the Epstein lab, and while we walk over there, we chat some more about navigation. I ask

Russell Epstein, head honcho of the navigation research lab here, if he is one of those scientists who went into a field of psychology to try and explain his own shortcomings. 'Actually, I'm pretty good [at navigation],' he replies. 'I've always been kind of interested in maps.' In fact, he has done a lot to map the brain, too. In 1998, as a young researcher fresh out of his PhD at Massachusetts Institute of Technology, he demonstrated that the brain has an area, called the parahippocampal place area (PPA), that is dedicated to processing scenes rather than objects. It does this, he found, by noticing that there is some kind of three-dimensional structure or geometry to what you are looking at. If you look at a photo of a room, for example, the PPA responds to that as a 'place', but if you cut out the images of the furniture and paste them onto a blank page, the PPA would be less excited about it.

The PPA is just one of the areas in my brain they are going to look at, along with the hippocampus (to see whether it's as shrivelled as I suspect it is) and the occipital place area (OPA), which is a fairly new addition to the navigational circuit diagram.

First is the structural scan — the bit where they measure my hippocampus and compare it to other women they have scanned in the past (there's no point comparing it to male brains; women's brains are smaller, yet perfectly formed ...). All I have to do for this part of the scan is lie still and try not to fall asleep. Then comes the functional scan, where they measure which bit of the brain is active while I look at whatever they flash on the screen in front of me. In this case, they show me a mixture of faces, objects, scenes, and scrambled-up images, which appear for a couple of seconds at a time. I later find out that this is so they can find each of the place-sensitive areas, since all three of them — the OPA, PPA, and RSC — respond more strongly to scenes than objects or faces.

This kind of imaging, it turns out, is even noisier than a structural scan — it makes a sound somewhere between a pneumatic drill and a fire alarm. Luckily, I have something to do to take my mind off it: I'm told to concentrate on the images and press a button if I see the same picture twice. Afterwards, they fess up that this was just a trick to make sure I stayed focused and didn't drift off. They needn't have bothered — I don't feel like attention is so much of an issue anymore. The butterfly seems to be more or less under control, either because of my meditation practice or perhaps because I'm having a fantastic time.

Just when I'm really starting to enjoy myself, there is one more test, which I later discover is to see how the same three areas respond to large things that might make a good landmark, over small things that are probably less useful. Cruelly, they tell me that there will be a memory test afterwards, so I am to remember as many objects as I can. Being a good girl, I do my best to memorise everything I see, only to find out that there wasn't a test, which means that I filled my memory banks with hundreds of images of staplers, wardrobes, and photocopiers for no reason whatsoever.

The next day comes the moment of truth: the big results reveal.

I'm impressed: the team has not only analysed a huge amount of data overnight, they have also knocked up a natty PowerPoint presentation to put them all together. As I suspected, all of the behavioural tests point towards me using an egocentric strategy: computing where to go based on where things are in relation to my body rather than any kind of mental map.

In fact, the data suggest that I am borderline incapable of forming mental maps and holding them in mind for a useful amount of time. In the grid test, for example, while I could recreate the view I had seen from any side of the table, I struggled to put the objects in the same position once I was facing another direction. To me, it felt like the information just slipped out of my head the second I had to use it, a bit like when my mobile phone auto-rotates when I'm in the middle of reading a sentence, and it takes a while to find where I was. It's disorientating, and very annoying.

Steve Marchette, who is a spectacularly nice guy, tried to make me feel better about the graph showing how I compared to other volunteers. 'You're still on the normal distribution [curve], but you're on the tail ...' Russell jumps in to correct him, 'She's pretty far out on the tail ...' And he's right — all the evidence points towards me being terrible at mentally shifting perspective.

The results from the three virtual-reality tasks also make a very clear case that I am fine at knowing where I'm going as long as I follow a route dotted with familiar landmarks. At no point in the maze task did I risk a shortcut; instead, I stuck to the exact routes I had learned previously. For example: from the traffic barrier turn right, and right again, and there is the bin. From there, go left, then left, and there is the chair. If I was told to collect the chair, I'd always take the long route, rather than risk getting lost by skipping from the barrier to the chair.

In the open-arena game, where there was no route to learn, only mountains around the outside and a small landmark (a ball) inside, I still used the one and only landmark rather than the huge, unchanging mountains all around me. I was aware while I was doing it that it would make much more sense to use the mountains, but I just couldn't seem to remember where they were in relation to each other and hold that information in mind. Again, my head kept spinning whenever I tried.

Using information about a large-scale scene rather than smaller landmarks is a hippocampal strategy, Russell tells me, and I'm clearly not very good at it. Which makes me wonder how my hippocampus is working — or not, as seems more likely.

I don't have to wait long to find out. The next slide shows the results of my brain scans, with the hippocampus outlined in red. Steve reassures me it's very shapely, which is nice to know. Then he hits me with the detail: if you break it down into different areas of the hippocampus, which are known to do different jobs, it starts to look more interesting. In fact, my brain is almost the exact opposite of what Eleanor Maguire found in the London taxi drivers.

In a 2000 study, Maguire found that taxi drivers who had been learning their trade for longer developed a larger posterior (back) section of the hippocampus, while the anterior (front) part of it got smaller.[5] In my case, this size pattern is reversed: the posterior part is in the bottom 10 per cent of size compared to other women they had measured, while the front is slightly larger than average. In Russell's studies, too, he tells me, people with a bigger posterior hippocampus on the right side of their brain tend to be better at forming cognitive maps. So if mine is small, it might explain why I am not so good at mental map-making.

Interestingly, while the posterior hippocampus is largely tied up with spatial memory, the anterior part has a role in processing anxiety,[6] and trait neuroticism has been linked to a generally smaller hippocampus. A study by Giuseppe Iaria's team at Calgary University showed that the higher trait-anxiety is, the worse a person's performance on spatial cognition tests.[7] So perhaps it makes sense that the emotional bit of the hippocampus would be larger than average in neurotic old me, and that I am bad at navigating. In that regard, I'm still doing the meditation and happy-faces clicking and feel like it's improving — but it's definitely a work in progress. Hopefully, this is all in the process of balancing out.

Later, when I send these results to Klaus Gramann back in Berlin, he is curious to find out if his results with small hippocampus / larger caudate nucleus applies to me as a user of landmarks rather than cognitive maps. I ask Steve if he'll measure it on my scan, which he very kindly does. Sadly, though, it turns out that I'm not particularly well endowed in that area either, particularly on the right side, which, like my right hippocampus, is way below average.

What does it mean? I ask Klaus. 'I have no idea what this actually implies, but definitely nothing that should make you value your orientation abilities less ...' he says, in an email. 'That's the reason I don't look at my brain scans.'

In case I needed any more evidence for a below-par navigational system, I get some from my functional MRI scan, which measures not size, but activity in key navigational areas. The good news is that of the three areas of the brain known to specifically care about scenes rather than anything else, two of them look pretty normal. The parahippocampal place area, which is Russell's baby, looks fine — as does the retrosplenial complex. The occipital place area, though ... not so much. 'We couldn't identify a very strong OPA for you,' Steve tells me, slightly apologetically.

But I'm delighted — at last, some evidence that my brain isn't working as it should. But what does it mean? Which bit of kit am I lacking? 'The OPA's function is not super well-known but Josh can say something about that ...' says Steve.

Josh, a PhD student in the lab, isn't telling though. 'I don't really want to speculate on it at this point ...' 'Right, but we have some reason to think that it might be specific for coding locations to boundaries?' Steve chips back in. Josh nods. 'Large-scale spatial information,' he confirms. Which, unsurprisingly, I'm not terribly good at. It looks to me like the reason it feels like I can't compute a large scene while I'm in it is because I physically can't — the bit of the brain that should be doing it is sleeping on the job. It's quite a relief, a bit like finding out that the reason I can't see without my glasses is because the lenses in my eyes are the wrong shape. Not my fault; just one of those things.

There might also be more to my normal-looking retrosplenial complex than meets the eye, too, Russell tells me. 'The RSC is active when you do these tasks that you're bad at, like imagining yourself in the location and figuring out where something else is. So you can say, "Oh you look great on that." But here's the interesting thing, there was a study done on a person with [a condition called] developmental topographical disorientation who gets totally lost all the time. And if you do this test, their RSC looks normal. But if you take it a step further and do more probing of function of this region, it starts to look a bit abnormal. And that's something that we didn't do with you.'

In other words, my RSC might react normally when it sees a picture of a place, but if you looked at how the RSC reacts when I imagine standing at one place I know well and point to another, it might not respond as much. 'Normally, the RSC would respond even more on that task — perhaps it wouldn't in you,' he suggests. The team did discuss doing this experiment with me, but decided that it would require several more hour-long sessions in the scanner, and even then the results wouldn't be that reliable. It's the same problem I keep coming up against: scientists never do what I am trying to do and understand how minds work one brain at a time. There are too many other factors that might explain the results on just one person, but if you scan tens or hundreds, you can notice patterns. Brain size measurements don't actually predict anything, statistically speaking, which is why we are nowhere near being able to scan a person's brain and use it to understand or predict their behaviour. On the other hand, like everything I have tried so far, the results do seem to line up remarkably well with my own experience. It might not be scientific-paper worthy, but it's fascinating nonetheless.

So the million-dollar question: if some of my circuitry isn't working properly, can I improve it with practice or do I need the navigational equivalent of contact lenses? Russell answers quickly, because I've been asking him this a lot over the past couple of days. 'We have no idea!' Steve is more hopeful, but gives more or less the same message. 'It's not that an argument from all the evidence says you can't get better, it's more that there is no evidence,' he says.

But what about the taxi drivers? They managed to grow their posterior hippocampus and become superhuman expert navigators. Can't I do something similar? Russell surprises me by saying that he isn't convinced that what the taxi drivers are learning are actually navigation skills. They are certainly getting better at learning routes around London streets, but is that improving their navigation ability or memory? It's a new one on me — I thought navigation was a done deal as far as brain changes go. I see what he means though: improving navigation would mean that London cabbies would be better at learning to navigate New York, LA, or anywhere else in the world, better than the average person. And that, he says, we don't know. 'That's an important question that we don't know the answer to.'

Yet again, the question of whether you can 'change your brain' — in the sense of rewiring circuits to make them work better — is actually an open question for many scientists. The folklore surrounding neuroplasticity might make it seem possible, but in some cases it might not be. If my OPA doesn't respond properly to scenes, for example, that might make it physically difficult to make a mental map. If that's the case, where do I even start in trying to train my mental-map-making skills? It'd be like trying to learn a language underwater, from the other end of the pool — the right messages just wouldn't get through no matter how hard I listened.

Another possibility, of course, is that if I practise navigating by calculating my distance from boundaries I might start to find it easier. One possibility is to start taking an interest in my son's computer games. Recent research found that novice gamers gained grey matter in their right hippocampus after two months of playing Super Mario Bros for 30 minutes a day.[8] Not only that, but they started to favour an allocentric strategy over an egocentric one as they became more expert at the game. Computer gaming has also been linked to improved ability to focus attention, which does make me wonder if a quick stint on Lego Star Wars with my son of an evening could cure a multitude of ills.

The only catch is that I have always hated playing computer games (probably because I am so bad at them), and in the Super Mario study, the degree of improvement, and grey matter increase, depended on how much the volunteers wanted to play video games. It's the same old problem — you need to pay attention to something to get better at it. You can't expect any brain change if you don't focus on the task at hand. I'm in two minds: the working-memory training made it very clear that life is too short to force yourself to do something you hate in the hope that it might improve your brain skills, but on the other hand the face-clicking has proven to be useful and worth the daily investment of time.

I ask Russell whether he thinks I should start playing video games to improve my spatial navigation skills. He answers, 'That depends if you want to play more video games!' Which is a good point. Maybe I should skip the games and find a way to add what support I need from elsewhere. A much better use of my time, and indeed brainpower, seems, to me anyway, to be to practise the actual skill that I want to improve — navigating in the real world — using very specific tools to support me while I learn. After all, I don't berate myself for not being able to reach a high shelf and try to train my body to get taller: I just grab the nearest chair and stand on it.

This is where the feelSpace comes in. My plan is to wear the belt for the prescribed six weeks, get out in the countryside, and see if it turns my random wanderings into a mental map of my hometown.

## FEEL YOUR WAY ...

Back at home, I treat my dog to six weeks' worth of long rambling walks in the countryside. Surprisingly, considering how self-conscious I feel, not that many people seemed to notice the belt, even though I decided early on that trying to hide it under clothes looks both ugly and dodgy, and have been wearing it loud and proud on the outside. Those who noticed and expressed an opinion guessed that it was either, a) a bum-bag for storing dog treats, b) one of those waist-toning devices, c) a TENS machine to relieve back-pain, or d) some kind of colostomy bag. Only one person picked up on the subtle buzzing, and said, 'You sound like you're having a nice time there ...' with a wink. Only once did someone shout 'Oi ya bomber!' out of a passing car, and it turned out to be someone I knew.

Fig 13. Me, my dog, and my best innocent face.

Incidentally, as we enter the woods on the first day, I'm reminded of a study I read about a while ago that said that dogs align themselves north to south when they poo.[9] This is the perfect opportunity to test that on my dog, Jango. And sure enough, more often than not, if I face the same way as him while he does a poo, the belt buzzes at my back. Actually, his compass seems to be a bit off — most of the time he seems to face either south-east or north-west. There must be a snappy way to describe this: bum to the sun? No, that isn't right ... Poo to the pole? I will keep an eye on this; as most of the navigating I am going to do will involve taking the dog out, there is plenty of opportunity to gather data.

The other thing I discover is that the path we enter the woods on faces roughly north-south. The loop I normally do is more or less square, with a few different paths heading off to the sides; I have walked there so often over the years, I have a pretty good mental map of it. The belt backs this up: the paths that I always assumed are going in the same direction as the first path into the woods actually are. Over time, I begin to put this information together for different areas: this path heads north, and my house faces north, so it must be over ... there.

As the weeks pass, I find that more and more landmarks in my town start to line up like this. I'm surprised to find that looking across the river in town I was facing north, the same direction as my house faces. If you'd asked me before, I would have guessed that my house faces in the same direction as the river flows, not looking across it. Over time, I can almost see familiar landmarks swivel and slide around in my mind's eye and click into place, all lining up to north. As a result, the whole layout of my hometown starts to make sense to me for the first time. I know that the nearest big town is north-east of here, and when I find myself driving into the sunset one evening I have a navigational 'aha' moment: 'The sun sets in the west — yep, that makes sense, because town is north-east of here, and it's behind me and over there.' I've even started noticing how the sun moves through the sky during the day, something that seemed like black magic before. 'Oh look, the sun is in the south at midday ...' It suddenly seems like a skill that might come in handy if I was ever lost in the wilderness.

I also discover when I get home that my wanderings with the belt in Berlin after I left Klaus' lab have left me with a pretty accurate mental map of central Berlin. One afternoon, my son is watching *Go Jetters* — a kid's TV programme about major landmarks. Today's episode is about the Brandenburg Gate (an evil baddie has stolen the horse and chariot on top of it). As soon as I see the Brandenburg Gate, I think: 'Ah, that's looking west. The Reichstag is north of there, and I walked right to get to it.' I never, ever think like that, normally. I couldn't tell you which direction Buckingham Palace is in relation to Piccadilly Circus in a million years, and I lived, and cycled, in London for years (I did a test of my mental map of London while I was in Philadelphia; you can see how London looks in my head below). Back in cartoon Berlin, when the Go Jetters' boss says that the baddie is heading east, I think about it for a second, watch the action, and think — actually, that's right! Checkpoint Charlie is sort of ... that way. I spent one day in Berlin with the belt and it has cemented the landmarks in my head in relation to their compass positions. That's pretty amazing.

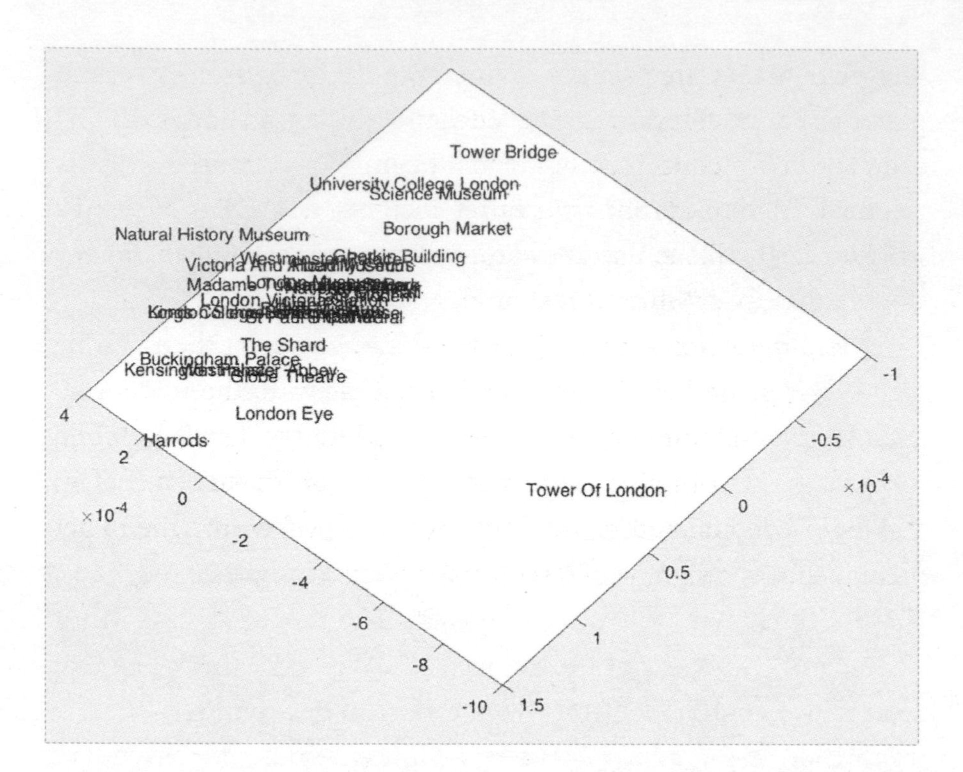


Fig 14. The map of London in my head as measured by the iJRD app. At least I know that the Tower of London is in the east ... Try it at: http://spatialcognitionapp.com/iJRD.html

It's not perfect, though, and in unfamiliar territory, knowing the position of north isn't enough, even with a prominent landmark for guidance. On one day, I decide to try walking on a huge area of heathland about 15 minutes from home. I start the walk by a large war memorial that is easy to see from large swathes of the heath, and set off north. My plan is to go south a bit, then right (west), and do a loop. Fast forward an hour or so, and by my calculations, any minute now I should see the memorial to the left (south-east). And ... it turns up to my right.

It would appear I've been walking in the right direction, but have gone too far. A compass is only useful if you also know how far you have gone in each direction, and landmarks are only useful if you know where they are in space. A line from one of Russell's research papers seems relevant here: 'Identification and localization/ orientation are conceptually distinct operations: a tourist in Paris might be able to identify the Eiffel Tower and Arc De Triomphe without being able to use that information to figure out where they were in the city or which direction they were facing.'[10]

One thing is for sure. If the war memorial wasn't there, I'd be lost — belt or no belt. In the end, I trudge back to the memorial, and turn around until I recognise my surroundings. I spend a while trying to figure out where I went wrong, fail to make sense of it, and then, disappointed, retrace my steps back to the car. When I get there, I spot this sign, which seems strangely appropriate.

After seven weeks with the belt, I reluctantly send it home to Germany. I've enjoyed playing with it, but has it changed anything about my brain's navigation circuitry? I have to be honest: I'm not sure. My brain scan in Philadelphia took place after my feelSpace adventure, and studies where brains were scanned before and after an intervention show that two months is long enough to start to see changes. I had the belt for seven weeks, in the end. So, by one token, if my brain was going to start looking like it could navigate properly, it probably would have by then. On the other hand, it has definitely changed the way I perceive my hometown, and now that I have the mental map in order, I have begun to use it to get around. Having the belt, and now a compass, also helped me to learn new navigational cues, like the position of the sun through the day, and this is something I can now use to navigate in a way I never would have before.

I contact Giuseppe Iaria at the University of Calgary again, because in addition to looking at how the hippocampus varies with navigation ability he is also developing training to improve navigational skills in people with developmental topographical disorientation (DTD). This is an inherited inability to form cognitive maps that can be so severe that people get lost even in their own home. His team at Calgary has online navigational tests and training programmes that anyone can try,[11] but which were offline being upgraded while I was working on my own skills. Giuseppe and I arrange a Skype meeting to discuss my results in relation to his work.

He agrees that it is a good idea to concentrate on building cognitive maps, because an egocentric strategy, while it kind of works, runs up against the problem of limited brain processing power, sooner or later. 'You can't remember an infinite number of turns,' he says. On the other hand, 'If you build cognitive maps you don't need to remember sequences, you don't need to remember specific landmarks, you don't need to pay attention all of the time. Focusing on creating a mental map is probably the best training that anyone can offer to anyone.'

In his research, this involves online training to try to get people with DTD to a level where they can build on their skills in reality.

For the rest of us, it seems that my hunch that the best training was to practise in the real world was right. 'If you are in a safe place, so you know nothing is going to happen, you can say — okay, I will go and play for one hour every day. It's a physical activity and physical activity is very important for the hippocampus. If you are able to do that for an hour a day and try to orient yourself before you use any tools to help you get home, that is about the best thing that you can do for yourself.'

Whether or not it will change anything, brain-wise, Giuseppe agrees with Russell that this is very much still an open question. 'We have so much knowledge about which parts of the brain are important for doing something. We have so much information about which cells are important in rats as they move around the environment. It's all great; it's fantastic. But we don't have a clear idea of how the brain is integrating all of this information in order to give you a sense of direction and navigation in large-scale environments.' That, Giuseppe says, is what his research is hoping to answer over the next few years.

More generally, what this tells us about the brain is that, while it does tend to get stuck on a certain way of doing something, such as navigating by landmarks rather than mental maps, it can change to a new strategy with a little bit of training. You just have to know that it's there. 'The majority of people who mention having problems with navigation and orientation, they are not problems with navigation and orientation; they are just set on a specific bias, they are not necessarily considering other methods. But as soon as you get that knowledge that you can achieve a goal by using a different approach, and obviously you need to practise because you are not used to using that approach, you will get better and better,' says Giuseppe.

That much is definitely true for me. When I started this project, my aim was to change my brain, and prove what had happened using before-and-after brain scans to track my progress. Now my focus is a little bit different: I want to change my abilities, whether that means adding new skills or using what I've got more efficiently. And given my newfound ability to think about my space in a different way, I think it's definitely possible.

More intriguingly, though, it seems as if I have stumbled upon another truth about the human brain: it is flexible enough to integrate a totally new sense. That is where the feelSpace project came from in the first place — the researchers wanted to know if people could integrate information that they have no sense organ for and incorporate it into their understanding of the world. It was only when volunteers started finding that it changed their perception of space that they thought of using it for that purpose.

I've got to be careful here, because I don't want to make it sound like I have sprouted a brand new bit of brain called the 'magnetic detection cortex'. Instead, what my brain has done is take a new and previously irrelevant bit of sensory information — a buzzing sensation on my waist — and linked that in memory to the position of north. Once that information was in memory, I could learn to tie it to landmarks. Now, it doesn't matter if I never wear the belt again — this information is there, intrinsically linked to my knowledge of my hometown. So whether or not I might have made any major changes to my brain's inbuilt circuitry, I have found that it is absolutely possible to bolt on a new skill that no human has evolved with naturally. And that, to me, is the most exciting way that I have 'changed' my brain so far.

*NAVIGATION TIPS FOR THE HABITUALLY LOST*

1. Get to know which part of the navigation puzzle may be lacking. Researchers at the University of Calgary have online tests you can try at www.gettinglost.ca.
2. Practise navigating in unfamiliar territory: get a map and compass, and switch off your phone and go exploring — you can always switch your phone back on again if you get really stuck. From experience: take snacks.
3. Learn to use the position of the sun as a makeshift compass, and pay attention to it when possible — a map might just form of its own accord.
4. Wait for haptic sensor belts to hit the market; the next generation of feelSpace, for example, will allow you to set the position of your goal, and it will buzz continuously in that direction to guide you in. Easy.
5. Consider fostering a gaming habit — it might just help.
6. Don't feel guilty about using Google Maps. Recent research suggests that having an aerial map with a regularly updated picture of north on it actually makes us better navigators.

CHAPTER 5

# Mind-bending, Time-bending

*'We must put aside the idea of a single time; all that counts are the multiple times that make up experience.'*

— HENRI BERGSON, *DURÉE ET SIMULTANÉITÉ*, 1922

*'People assume that time is a strict progression of cause to effect, but actually ... it is more like a big ball of wibbly-wobbly, timey-wimey ... stuff.'*

— THE TENTH DOCTOR (*DR WHO*)

I'm no physicist, and indeed no Time Lord, so I am not going to try and explain what time actually is, in the real, physical world; I'm not even sure that it's possible (for what it's worth, my physicist friends confirmed that I'm about right on that). But you don't need to be a physicist to know that our experience of time passing is anything but constant.

When I crashed my car, head on, 15 years ago, I discovered just how slowly it can go. In the seconds before I hit the car coming the other way, it was like being in a slow-mo movie sequence. I remember turning the steering wheel right and left for what felt like ten seconds in a vain attempt to miss the other car. I had several distinct thoughts while this was happening: 'Oh god, I'm going to crash. I'm only 26 and I'm going to die. This must have been what it was like for Dad.' Then time rushed forward at double speed as I hurtled towards the other car, before snapping back to normal speed: *BANG*. *Ouch*. Silence.

My wedding day was the complete opposite. From getting my hair done in the morning to the last dance at midnight, it went by in a complete blur. It all whooshed by so quickly that I barely remember the details of the day — who I spoke to and what we said. I only remember the details of the vows we made because there is video evidence of what we agreed to.

This wibbly-wobbliness of time is well known, and is understood to be a product of how our brains process what is going on around us at any given moment. What I want to know is: is it possible to learn how to manipulate time at will, not as a passive observer but as the driver of your own perception?

Controlling perception of time is something that is often written about in popular science articles on the subject, and it sounds pretty appealing — particularly to a 40-something like me who would like to know what the hell happened to the last decade. Annoyingly, though, I'm yet to find any useable advice on how to actually do it in real life.

The standard spiel is to fill your time with new and exciting experiences. The explanation is that, when you are a child and summer holidays last forever, it is because everything is shiny and new and worthy of your full attention, and so you are constantly storing new things in memory. When you look back at that time, you get the mental illusion that it must have taken ages to fit in all that exciting new stuff.

Once you're an adult and life slips into a kind of predictable rhythm of work, socialising, and maybe a couple of hobbies to break up the chores, we start to live life more or less in our sleep, hardly bothering to pay attention because we've seen it all before. Fewer attention-grabbing experiences means that a smaller number of new memories are being stored — so when you look back on your recent past, it feels like it must have gone quickly. It's all a bit sad when you think about it: if our lives are the sum total of our memories and experiences, and we aren't bothering to have any, are we really having a life at all?

But I have a problem with the 'get a more exciting life' solution to stopping life speeding past. It's just not very realistic. It reminds me of the glib advice to 'live every day as if it were your last'. There's nothing wrong with the sentiment, but it's not the way the world works. Most of us need to work for a living, usually for most of the hours in the day — which leaves barely enough time for the boring stuff, let alone any swinging from chandeliers.

More than that, though, adding new and exciting things to your life only allows the illusion of control over how quickly time passed when you look back at the experience afterwards. What I'm interested in is making time pass at a different speed as you are actually living in it, and that's not necessarily the same thing.

As it turns out, this distinction between time judgements in the moment and after the event is a topic of debate in time-psychology at the moment, led by veteran time-researcher John Wearden, of Keele University, in the UK. Wearden has been working on time perception since the 1980s, and — along with Sylvie Droit-Volet of Clermont University, in France — has recently published new research showing that, while many researchers take time judgements made after the event to be a good proxy for how time felt in the moment, the two measurements often bear little relation to each other.

In one of their recent experiments, volunteers were interrupted by a smartphone app that asked what they were doing, how they felt, and how quickly time was passing in the moment. They were also asked to either hold down a button on the phone for a certain duration — 500, 1000, or 15,000 seconds — or to estimate how long a stimulus on the screen appeared for. These are all different ways of measuring a person's accuracy in estimating time as it is passing and after the event. 'What we found, basically, was the duration judgements were completely disconnected from judgements of the passage of time,' Wearden says. 'If you reported that time was going quickly, it didn't affect your judgement of [how long] a second [lasts]. The two seem be based on completely different things.'[1]

This chimes with what he found back in 2005: people watching an exciting movie feel like time is whizzing by in the moment, but if you ask them afterwards to guess how long the clip lasted, they guessed that they had been watching it for longer than they really had. The opposite was true for people who watched a relaxation film. Which means that you might feel that time is going quickly in the moment, but afterwards feel that it took ages. So which is the 'real' perception of time? In this case, did the exciting movie take more or less time? Both, is seems, are illusions created by the brain, and making that judgement depends on which illusion you choose to believe at any given moment.

I arrange a Skype chat with Wearden to see if he can throw any light on the subject. I have interviewed him before, and remember him as the kind of charmingly grumpy guru who is great fun to chat to because he's knowledgeable enough to know every twist and turn in the past 30 years of research in time perception, and forthright enough to tell me when his opinion on a particular theory is that it's 'a load of complete bollocks'. (Scientists' usual code for this is: 'It's an interesting idea, but there are a number of problems with it ...')

Wearden tells me that the trouble with what I am trying to do is that in-the-moment time hasn't been particularly well studied. There has been, he says, 'a lot of speculation about it and a little bit of research.' There has been no shortage of perception-of-time studies, of course, but most of them use duration judgements, given after the event: 'how long did that movie clip last?', 'how long have we been talking?' or prospective judgements, made in advance, such as 'hold that button down for three seconds'.

A handful of people are tackling in-the-moment time perception, though, which is good news for me. I'm not bothered about feeling as if a two-week holiday lasted forever after the event — I want to control how I perceive time in the moment. It would be the ultimate use of the mental flexibility I've been trying to foster so far. Imagine if you could choose to remember every second of walking down the aisle and saying your vows as if it was in the slow-mo of a car crash, every thought and feeling larger than life and twice as colourful? And wouldn't it be great, when you can feel yourself getting bored and time dragging to a standstill, to be able to do something specific to speed things along?

Unfortunately, there are one or two other reasons to think that this isn't going to be easy. What I'm basically trying to do is get a grip on the nature of human consciousness: that mysterious sense that there is a 'me' that is living in the flow of another mysterious thing called time. Far cleverer people than me have been flummoxed by both of these questions over the course of human history.

Not only that, I'm not only trying to understand consciousness: I'm attempting to change the very nature of that consciousness as is it happening to me. Essentially, I'm trying to choose one illusion over a different illusion while simultaneously living *in* an illusion. Maybe the second you stop living in time, and turn your focus onto it, it will change the experiencing altogether. Even more confusingly, time isn't a solid immovable 'thing' that is perceived in the brain at all, at least not in the same way that the brain perceives a table or chair. Each person's perception of time is generated in the same brain that is perceiving the rest of the experience. Which makes me wonder whether it is even possible for anyone to understand their conscious experience of time from the inside.

All in all, thinking about time is making me feel dizzy in the same sort of way as when I look at the night sky and try to get my head around our place in the universe. Intellectually, it all kind of makes sense, but if you try and put yourself and your life into that picture, it very quickly blows your mind. It also makes me realise how much we take for granted the passage of time as a framework for making sense of life; it plays a huge part in our lives and we don't have to do anything to make it happen.

Not everyone has this luxury, though, as I discovered by accident while talking to a dog-walking friend one day. I've known Jeannie for years — we have very similar naughty sheepdogs, who love to herd each other into tiny circles, while arguing over which one of them is the sheep. Jeannie's a lovely woman, eloquent and softly spoken but with an impish sense of humour that erupts into a loud explosive laugh when you least expect it. I had noticed that she sometimes seems a little distracted, but didn't think much of it; I'm sure I come across like that too, when I'm having one of my more creative, 'hypofrontal', moments. It turns out, though, that the reason she seems distracted is that since suffering a stroke six years ago, she is totally lost in time. Without regularly checking her watch, she has no idea whether we have been walking around our local park for ten minutes or an hour, and without consulting her to-do list she wouldn't know whether the next thing she has to do is go home and cook the dinner, or to go to work to start the day. In fact, more often than not, she's not sure what day of the week it is, what season we are in, and sometimes even where she is on the timeline of her own life. It's something, she tells me, called 'loss of passage of time syndrome' and it sounds both terrible and fascinating.

Over a cuppa at Jeannie's house a few months later, I get the full story. She first noticed the problem a few weeks after leaving hospital, as she started to ease back into normal life. 'I'd get up in the morning to get my sons off to school, and my husband would come up and say, "Come out of the shower!" I had no idea whether I'd been in there a minute or, as it happened, 40 minutes.'

Six years on, this is the kind of thing that still happens all the time. She tells me about a recent occasion where she nipped over to check the notice board at work and the next thing she knew a colleague had come to find her because she'd been missing for 20 minutes and was needed elsewhere. 'How long did it feel like you'd been there?' I ask. 'Five minutes?' 'Less,' she replies.

This is particularly interesting, because it is often assumed that the ultimate state of wellbeing is to be in the flow, to step out of time and ride the waves of mental freedom. From what Jeannie tells me, though, it's not a state that she would recommend long term.

'It's just this ghastly feeling of untetheredness,' she says. 'We all have that ticking clock that we touch base with — and for me there is something that drags me away, and I don't know where, but some terrifying place that isn't to do with time. And there's a certain feeling that comes with that, that isn't freeing or pleasant, it's sort of a dull dread ... It's like a lostness on a hundred different levels.'

Listening to Jeannie talk about the unbearable lack of time, it is obvious that being able to check in with some kind of ticking clock is crucial for our emotional and cognitive wellbeing. Most psychologists agree that there probably is such a thing ticking away somewhere in the brain. They know this because experiments in animals have shown that if you feed them by the clock, all kinds of creatures, from fish to rats to turtles and dogs, are able to count time well enough to know when to expect dinner to appear, and to look a bit disgruntled when it doesn't. Only humans, psychologists presume, experience time flowing in the moment. Unless, of course, any other species are truly conscious — which they might be, but that is a whole other story.

The leading model that explains the internal clock of this is called the pacemaker-accumulator model. In this view, the 'in the moment' bit of this is guided by attention, which acts as a kind of switch that turns on the ticks of the internal clock. These ticks are temporarily stored in an 'accumulator', a kind of holding pen where they are temporarily stored to be counted if needs be.

When we want to work out how long something has lasted, we use memory to access the accumulator and count the ticks. When we access the accumulator, it uses memory, not attention, to compare the most recent batch of time with our memory of past time spans. Wherever the internal clock is located in the body, more than one cognitive process is driving it. Which makes me think that you can probably only choose to tinker with one part of the machine at a time.

Psychologists know that this is a psychological clock, inside our heads, rather than any physical measurement of time, because changing what is going on in the body and mind tampers with the speed of the ticks. Heating the body or head (believe it or not some experiments used heated helmets to specifically warm the head), and taking drugs like amphetamines or anything that increases dopamine levels in the brain, make the clock tick faster — so that estimates of how long a stimulus lasted start to get out of kilter with clock time. Strong emotions, like fear and anger, do something similar, which is thought to account for the kind of experience I had when I crashed the car. When strong drugs and strong emotions affect the workings of the brain, our perception of time starts to go all over the place.

This is where it gets a bit difficult to visualise, because a faster ticking clock doesn't only mean that you think that time has passed faster in the moment. Instead, it makes you feel like *more* time has passed: when you count the ticks, there are more of them in there compared with what the real clock says. If there are more ticks, there must have been more time, so in retrospect you perceive that it has passed more slowly.

It makes sense, and nicely explains why time is so wibbly-wobbly in psychological experiments, but if you want to make a sweeping statement about whether time goes faster or slower in a given situation, you can easily come up with two different answers. When watching an exciting movie, the ticks go faster, so time seems to fly — but when you look back on it, more ticks makes you overestimate the time you spent watching it. Similarly, it's hard to predict whether, if I do manage to slow down the ticking of my internal clock, that will make time slow down or speed up; it sounds like it would make it slow down, but if there are fewer ticks in the accumulator, then that must mean it has seemingly taken less time, which should mean time has gone faster. It's mind-boggling. 'You are not the only one who is confused,' says Wearden.

In an attempt to make sense of it all, I decided to keep track of my own experience of time during specific 'moments' that seem particularly relevant based on what I have read so far. I want to see if any patterns emerge based on how I am feeling and what I am focused on. So, with advice from Wearden and Dan Zakay — another veteran time-psychologist, who is now at the Interdisciplinary Center Herzliya, a private research centre in Israel — I design my own experiment.

Much as I have enjoyed being locked in windowless rooms and zapped in the head, I didn't see any benefit of doing the experiments for this chapter in the lab. Lab experiments to measure passage of time involve getting people to do something, and then asking them:

is time going at clock speed, faster than the clock, or slower than the clock? I figure this is something I can do just as well by just living it and noticing what is happening to time.

I start logging my subjective experience of time (slow, fast, or the same as the clock), as well as making a duration judgement about an event afterwards. To do this, I set up the stopwatch on my phone, turned it facedown on the table at the start of a moment, then turned it off at the end without looking at it. Then I compared my estimate of how long I'd been doing something to what the clock said. The results are in table 3, below.

| | Guess time passed | Actual time passed | Over/under estimation | Time felt | I felt |
|---|---|---|---|---|---|
| Silence over dinner, after telling my son off | 6 mins | 3 mins | Over | Slow (bad) | Angry |
| Playing Lego with my son (and enjoying it) | 15 mins | 32 mins | Under | Slow (good) | Relaxed and engaged |
| Faffing about on the internet (should be working) | 15 mins | 5 mins, 15 secs | Over | Slow (bad) | Bored, annoyed with myself |
| Silent lunch (meditation retreat) | 20 mins | 40 mins | Under | Slow (good) | Relaxed |
| Roller disco | 20 mins | 40 mins | Under | Fast (good) | Happy, having fun |
| Breath and body meditation (part of day retreat) | 30 mins | 40 mins | Under | Fast (good) | Focused |
| Being in a hurry (school morning) | 20 mins | 30 mins | Under | Fast (bad) | Panicked |
| 10-hour trans-Atlantic flight (engrossed in work, followed by a film) | 5 hours | 8 hours | Under | Fast (good) | Engaged / entertained |

**Table 3. My estimations of time in the moment and after the event, and how it felt in the moment.**

What I found agreed with John Wearden's findings: my judgement of time in the moment, whether it felt fast or slow, didn't always match up with my judgement of duration afterwards. Sometimes, like when playing Lego with my son, time ran quickly, and I guessed that far less time had passed than actually had. But in the moment, time felt slow, because I was super-engaged with what we were doing. Other times, I was painfully aware of time — like that awkward silence at the dinner table after telling my son off — and my estimate about the amount of time that had passed was double the real amount. This felt slow, too, but in a bad way. Similarly, underestimating time was sometimes a good thing and sometimes a bad thing, depending on whether time was speeding along because I was having a great time or because I was frantically racing against the clock.

It's all very confusing — and to make matters worse, Zakay sounds a word of warning about reading too much into these results. They are, he says, anything but scientific. 'You know in advance what you are doing and what the expected outcomes should be,' he says. 'For example, if you know that duration is expected to be felt as longer under condition X than under condition Y, then you can press the stopwatch after a longer period under X than under Y, not because this is what happened with time but because you knew, maybe unconsciously, that this is what should happen.'

This is another basic feature of psychological research — the person being tested isn't supposed to know anything about what they are being measured on. The power of expectation can be enough, even unconsciously, to totally skew the results. Which is fine, but if anyone is going to get any practical use out of any of this research outside of the lab, at some point they are going to have to do what I did and give it a go, bias or no bias.

But the results are confusing and, frustratingly, reveal no obvious rules about specific mental states and their effect on the passage of time. Slow time can sometimes feel good and sometimes bad, and negative emotions can make time seem to run faster or slower.

If it's this difficult to measure my perception of time, how on earth am I going to change it? With no clear answers yet, I decide to follow my impulsive streak and just get stuck in with applying a few tricks that I have gleaned from various pieces of time-perception research.

## *TIME OVERRIDE EXPERIMENT 1*
## *MIGRAINE DAY*

Wearden's most recent study confirms what most of us already know: that time seems to pass more quickly if you are feeling happy, and more slowly if you are feeling sad. Whether you can engineer this kind of situation is less clear — I have my doubts, but am willing to give it a go all the same.

The perfect opportunity came along the day after my Skype chat with Wearden. For me, regular migraines are among the hormonal joys of womanhood — and although I hate them, on the plus side, each attack is predictable in that one is very much the same as the next. The moment I feel it creeping up from the base of my neck, I can guarantee that the next 36–48 hours are going to pass in a wonky-headed, blurry-eyed haze. And particularly in the hour when the painkillers have worn off but it's too early to take the next lot, time will drag painfully slowly.

To try and make this one pass a bit more quickly, first, I'm going to try to improve my mood. I don't have high hopes: I'm about as grumpy as a journalist with a sore head and a lot of work to do. But in the name of science, I download my favourite comedy podcast: the Adam and Joe show. These two always crack me up, so if anything is going to improve my mood, it's them. I turn the volume down as low as possible so it doesn't hurt too much, and hope for the best.

And ... it definitely cheered me up a bit, but dealing with the noise did make it feel like longer than the 20 minutes I managed to tune in for; their particular brand of silly song-based humour is a bit much for a migraine day, it turns out. My other option seems even more of a non-starter: find something that demands all of my attention, emotional investment, or a lot of skill. My head hurts too much, and my eyes aren't working together — so I can't even focus, let alone concentrate. Which leaves me only one time-bending option: give up altogether and go to sleep. I would love to do this, but I have to do the school run now. So, off I trudge, with one eye closed and a hat pulled over my face.

The verdict? Migraine: 1; time override: 0.

While wondering what on earth to do next, I happened upon a new theory of time perception, which specifically focuses on changing how time feels in the moment. It seemed perfect for what I was trying to do. Marc Wittmann, a psychologist at the Institute for Frontier Areas of Psychology and Mental Health in Freiburg, Germany, has come up with an idea, broadly called 'body time' — which sounds altogether more useable than an invisible internal clock somewhere in the brain.

Over Skype, from his office in Freiburg, he tells me that time is always, always to do with how we focus. I kind of suspected that already — in fact, given its central role in sustaining attention, anxiety, navigation, and creativity, I am starting to think that everything is about focus. But, he tells me, it's about more than just where in the outside world we place our attention. Much more important for our sense of time is where our focus is physically, inside the body, at any one moment. In fact, he goes as far as to say that time *is* what is going on in our bodies at any one moment.

'It's meaningless to say I'm attending to time — psychologists use this term a lot, but where are you attending to? My idea is that you are attending to yourself, to your bodily self, your mental self — and that is how you attend to time,' he says.

Wittmann has even suggested a good candidate for these kinds of 'in the moment' time judgements in the brain: an area called the insular cortex, or insula. If you want to imagine where this bit of the brain is, go about an inch above your ear on either side, and imagine pulling apart the top layer of the wrinkly cortex in the deepest fold there. Underneath is another layer of cortex, which is the insular cortex. The brain has a matching pair, one on each side of the brain.

The insula is a part of the brain that keeps track of bodily sensations, and processes emotions. These two things together are what gives us the impression that we are one person, made up of a seamless physical and emotional self. It also, he suggests, gives us the impression of a 'self' that is moving through time. If you wanted a glib tagline for this, you could say that the insula puts the 'me' into 'time'.

Rather than being a competitor with the internal-clock model, the body-time idea actually fits into the theory pretty well. 'You could say that the pulses are our body signals,' says Wittmann. As for the attentional switch — that is, the extent to which we are or are not paying attention to our body signals — this also fits.

Wittmann also has a slightly different but related line of work, which concerns how long a moment lasts in conscious awareness. Though our minds make a seemingly seamless story of our lives as we are living them, in reality the present moment only lasts an instant before being shunted into memory or forgotten altogether. If what I want to do is stretch that moment or ignore it until it goes away, then it would be helpful to know what I'm working with.

The general consensus is that a psychological moment lasts approximately two to three seconds. This was confirmed a couple of years ago in ingenious experiments where psychologists scrambled tiny fragments of movie clips to see if people noticed. They only noticed, and lost the plot of the scene, if the mixed-up scenes lasted more than two to three seconds.[2]

Intriguingly, this short window of now-ness feeds into our lives in all kinds of ways. Greetings and goodbyes — including hugs, kisses, waves, and handshakes — all last about three seconds on average. Holding any of these for much longer than that, especially with a stranger, is a sure-fire way to make both of you feel awkward.

More intriguingly still, Wittmann has done some experiments that suggest that, with enough training, it might be possible to extend this moment: he asked one group of expert meditators, and another of people with no meditation experience, to look at an optical-illusion image that can flip between two interpretations — the most famous is the Necker cube (below) which can be seen either with the cube going up and away from you towards the right or down and towards the left. Normally, the image flips between one and the other interpretation every two to three seconds, Wittmann says.

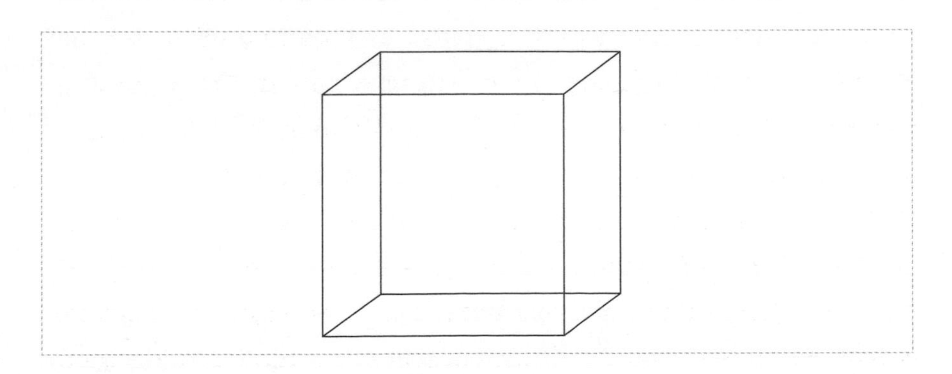

Fig 15. A Necker cube.

When both groups were asked to hold one interpretation in mind for as long as possible, the meditators were able to hold it for around eight seconds, two more on average than the controls.

Extending a moment by deliberately focusing your attention on your body or some aspect of your mental experience sounds like a pretty user-friendly way to slow down time as you are living it. It comes back to my mindfulness training: focusing on the minutiae of physical or mental experience takes more time than just flitting along in a dream.

It is not a tool for all occasions, though, as I found out to my cost. After the end of my eight-week mindfulness meditation course, I signed up to a full-day retreat with my teacher Gill. I had been pleasantly surprised by how much I had enjoyed the two-hour session on the last day, and wanted to know what a full seven hours of meditation practice would feel like.

As it happened, the day retreat fell on day two of yet another migraine. I was hoping that a day of relaxed breathing would do me good, and maybe that I'd even find a way to be 'present with the pain' so it wouldn't matter so much. Instead, it turned into possibly the longest day of my life, with the notable exception of the two that I spent in hospital having my son. Every time Gill asked us to 'notice what's here' I would notice that my head hurt, my neck hurt, and that I felt sick. Every time she invited us to notice what the body needs, I thought 'it needs to get onto that sofa over there and go to sleep'. By the end of the day, I was so frustrated I felt ready to explode, and spent the last half an hour of the practice biting my tongue so I didn't shout 'just ring the bloody bell!!'

I'm reliably informed that mindfulness can and does help people manage pain, with practice. In experiments, two groups of people were given the same painful stimulus (something hot against their skin), but one group was also tricked into thinking that time was passing faster than it actually was. These people reported experiencing less severe pain than the other group, for whom the clock was set at a normal speed.[3] Whether it is possible to increase the speed of the *internal* clock to get a similar effect is less clear. I couldn't find any research on the subject — and my 'Adam and Joe' experiment would suggest maybe not.

My experience of being mindful for a whole day did indicate that making each painful, nauseous moment feel like an eternity is a very bad idea indeed. When I told Wittmann about my experience, he looked horrified at the thought. 'You shouldn't do that, because you are so self-aware and so aware of your migraine. Pain is the perfect time modulator — time totally expands when you are in pain.' So I was in pain — which stretched time — then I stretched it further by focusing on my inner experience for seven hours. Good idea, Caroline.

What I take from this is that practising mindfulness might not be the best idea, if you are already struggling to get through the day. On the other hand, if you happen to be having a good time then choosing to be mindful of what you see, hear, think, and feel is a great way to potentially stretch the moment.

What about speeding time up to skip to the end of a bad day or a long and boring journey? This, says John Wearden, is altogether more problematic, because when someone says that 'time flies', they are nearly always talking about time that is already in the past. 'They say "I was so engrossed in this book, I looked at the clock and it was 10 o'clock. Therefore, time must have gone quickly." But they didn't *feel* it go quickly, they didn't really feel it at all.'

It's a problem that Wearden has begun to raise with his colleagues in recent years, many of whom don't seem to have thought of it before. 'I spoke to a few time psychologists and they said, "Yeah, now that you mention it, has anyone ever experienced fast time?" And, in a sense, you can't really, because you can't fast-forward reality.'

Which makes it seem that any attempt to skip to the end of a bad situation is unlikely to work, at least in the moment. One possible way around it perhaps, is to disconnect from the whole experience of time by going to sleep or by finding something that grabs your attention so completely that you lose track of time. One of the best things for this is television, Wittmann suggests. If all else fails, you might not be able to make time speed up, but you can almost certainly lose track of it. That's pretty much the opposite of the kind of mind control I am aiming for, however.

An alternative, according to Wittmann's theory of 'body time', is to do something to speed up the signals coming from the body. If the insula is constantly deciding how fast time is going based on signals from the body, perhaps exercise is a way to speed time along when you are bored? He agrees that it is definitely a factor. 'There are two things: one is attention to time — meaning attention to your bodily self — and the other thing is activation. Activating your body and feeling your body are other ways to affect your perception of time.'

First, this would make time speed up, while you are actually doing the exercise, he suggests — and then afterwards, the continued attention to body signals might re-set you to some kind of on-the-clock calm. 'Say I go jogging for an hour and then I calm down, but still feel very active; I am very much feeling myself and I feel myself in time, and I feel that everything is happening much slower because my body is much more activated and I feel myself much more intensely.'

To try out the slowing and speeding up of time based on activity in the mind versus the body, I arrange for a Saturday of intense time-perception self-experimentation in which I will experience two extremes in the same day. First, I will attend another silent meditation day retreat — this time, thankfully, without a migraine. Then, by way of contrast, my family and I will head to a roller-disco for the evening. Will a few hours of mindless exercise, high-energy pop, and slight anxiety about my son breaking a bone feel longer or shorter than hours of sitting, lying, and moving meditation where the only thing that moves is my attention?

Unsurprisingly, the roller disco flew by. The first 40 minutes felt more like 20, and I nearly forgot to look at the clock, because I was having such a blast, whizzing around to loud music and pretending I was 13 again. This all fits in to the body-time theory quite nicely: I was exercising, so my body signals were ticking by faster, and I was engrossed in the music, which took my attention away from time and let it fly by unnoticed. This seemed to work great, even though there was a huge digital clock on the wall and I was also aware that my son was up way past his bedtime and that his next fall could trigger a meltdown.

As for the meditation, my experience of time was mixed. If I took an average of the day, I'd say that time felt slow — although the first session of the day was the complete opposite and felt a lot shorter than 40 minutes; interestingly, that one was a meditation about bodily awareness.

## GO WITH THE FLOW

So that is slowing time and speeding time. Given the challenge I set myself in trying to change conscious perception while also living in it, I think it's about as much success as anyone can expect. I won't go as far as to say that I have applied neuroplasticity and changed my brain to perceive time differently, but I would definitely say that I have gained a few tips on how to use it more effectively — and perhaps more flexibly. So in that sense, it has been a success.

There is another state of psychological time, though, when time disappears and we step into another zone altogether. It's back to the idea of flow, or 'the zone' again, which I first played with in Boston, although at the time I didn't pay much attention to what it did to my perception of time. By now, I have over a year of practising that zone under my belt, and I'm intrigued to see what time-psychologists have to say about other potential ways to slip into the flow, and out of time.

John Wearden told me about the little-known research of Elizabeth Larson, an occupational therapist at the University of Wisconsin–Madison. She has been studying time perception from the point of view of patients who are injured or disabled. She figures that if you can work out what kinds of experiences get people into an enjoyable state of flow, then it might be possible to tweak certain aspects of patients' lives to make them more enjoyable and less stressful. If there are things they could be doing — things to get into a state where time 'flies by' or ceases to matter — this could be a good goal to aim for to improve wellbeing. It seems like something that we could all benefit from.

Larson has found that our perception of time shifts in a kind of wonky bell-shape (see graph below), depending on how well the needs of a situation match our skills. If whatever we are doing is so easy that we can do it without thinking too much, then time feels slow or about the same speed as it says on the clock. It speeds up and reaches flow when we are doing something that is just right for our skills and we are in the 'relaxed and ready', flow-like state, which I experienced (eventually) in Boston when the computer training was aligned to my own personal level. Then, when the job gets beyond our capabilities, we are unceremoniously ejected from the flow, and time starts to grind to a halt again.

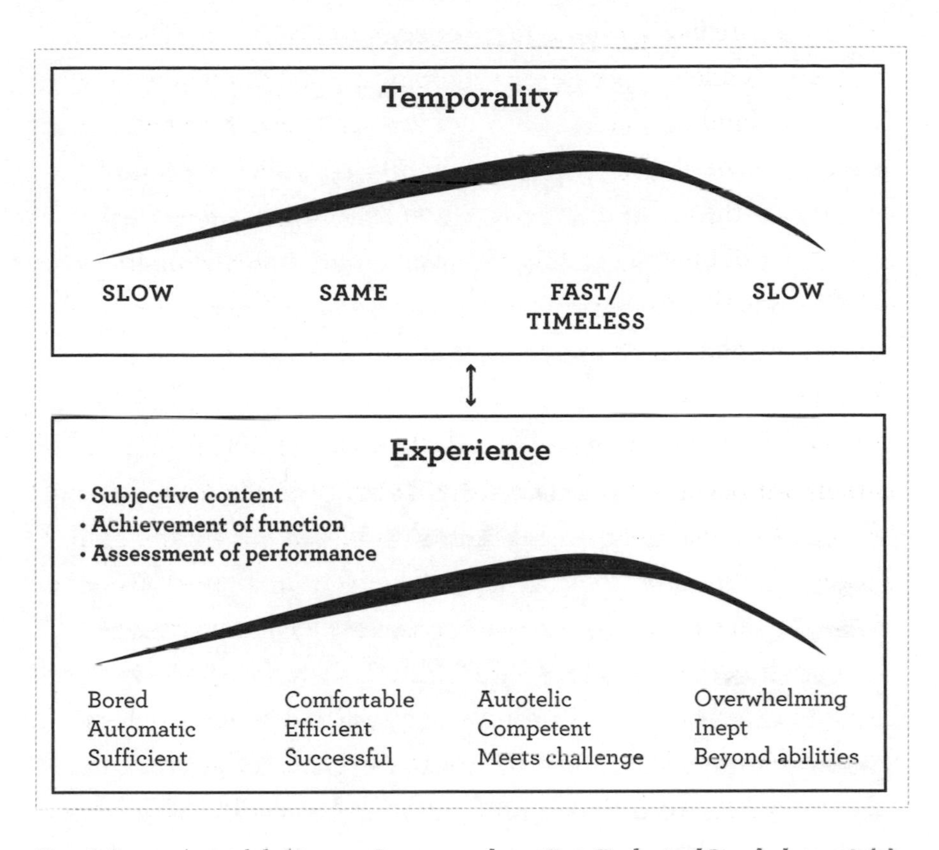


**Fig 16. Larson's model** (Source: Larson and von Eye, *Ecological Psychology*, 18 (2), pp. 113–30)

Which reminds me of my second time-bending experience — the long flight that wasn't: how does it feel to be trapped in a metal tube with no internet signal for most of my waking hours?

## *TIME OVERRIDE EXPERIMENT 2*
## *10-HOUR FLIGHT, LONDON–ATLANTA*

When I booked my trip to take in the labs of Russell Epstein in Philadelphia and Lila Chrysikou in Kansas, it made perfect sense to go via Atlanta. Then I looked on the map and the flight-time: ten hours, starting at 9.30 am, with a three-hour layover in Atlanta. It was going to be a long day.

On the plus-side, this was the first time for weeks that I had had a whole day to myself. No dog to walk, no dinner to plan, nobody wanting me to chat about their day or play Lego with them. All I had to do for the entire flight was get on with the work I'd been struggling to fit in recently. If I got tired of that, then I could ask the nice lady to bring me a glass of wine and I could sit back and watch a film. Bliss.

To anyone with a full-time job and/or no kids, it might sound strange to be excited about a whole day of work. For me, though, it's a much-needed chance to get my teeth into something that I find fascinating and just the right amount of mind-boggling. In short, when I'm in the mood for it, work puts me right into the top of Larson's wonky bell-shaped curve. That chimes with a study by psychologist Regina Conti, from 2001, which found that it matters whether you are doing something for yourself or just because you have been told to.[4] Only choosing to do a task for yourself sets you free from the drudgery of clock-watching.

And guess what? The journey flew by. I wrote 3000 words (some of which I didn't delete later) and enjoyed all eight of the hours I worked. In fact, it made me wonder whether internet-free travel might be the productivity tool I have been looking for. After eight hours, I had done everything I wanted to do, so I put the computer away, ordered a Baileys, and watched a film.

Before I knew it, we were landing in Atlanta.

So it does indeed seem to be true that if you can get your brain to engage properly, not only will it mean you are less aware of time in the moment, but after the event, it seems like less has passed. Even looking back on it after a long day's travel, the London–Atlanta stretch didn't feel that bad at all. It's a win-win.

Just don't ask me about the three-hour layover. By then, I was too tired to work, and there was only American football on the overhead screens, in a domestic terminal with no shops. Three hours feeling bored and tired felt way longer than ten hours of satisfying work. And nothing I did to try and change it made the slightest bit of difference. Damn.

Verdict? Long flight: 0; time override: 1.

Layover: 1; time override: 0.

So it seems as if Larson is right: one way to step out of time is to become totally absorbed in a task. But what if you have succumbed to boredom and no longer have the mental energy to drag yourself into that state? Here, it might be a better idea to revisit Wittmann's ideas about body time. If perception of time relies on physical and emotional signals being processed in the insula, what happens if you deprive it of any stimulation at all? Or perhaps flood the senses so that the body signals don't get a look in?

Depriving the brain of bodily and emotional stimulation might explain the timeless state that expert meditators describe once they have managed to quiet down all the mind-wandering. Another option perhaps is to become absorbed by music. Drowning out your body signals and playing with your emotions with the right tune might be one way to get into this state without that much effort.

It's speculative at best, but Wittmann seems convinced, so it's good enough for me: 'Totally,' he agrees when I suggest it. 'It's like instant meditation.'

Wittmann also describes the feeling of slipping out of time when he went into a flotation tank — a stress-relieving tool found in some expensive spas and dedicated flotation centres. The water in a flotation tank, or pod, is set to body temperature, and is salty enough to entirely support the body. Because you can't tell where your body starts and ends, and it is fully supported by the water, you drift into a state of calm that makes the stresses of normal life disappear.

'I went in there for two hours ... Your mind is wandering, and then suddenly it stops. I had an extreme experience of which meditators have, just after the second time of going in there.' With his body supported and nothing left to think about, he says, 'Time evaporated.'

I had a similar experience in a flotation tank many years ago, although it took ages to 'let go' of my tense neck, and let the water support my head. But I can kind of see what he means. It was a bit like that dreamy state before you fall asleep at night. I felt weightless, calm, and with no perception that time was passing at all.

So what can we take from all of this? For me, learning to override my natural perception of time has been a partial success. I may not have found the mental throttle that will allow me to speed away from a boring or painful situation, but that's probably because there isn't one. Time perception seems to arise out of whatever is happening in the brain and body right now — and if the mind and body are feeling tense and trapped in time, the only way out of that is to persuade either the body or the mind that it isn't. Some strategies that seem to work include: exercise, music, or, if you're not already too miserable, something engaging to take your mind off time.

On the other hand, slowing down the good times is a matter of learning to engage your attention to the minute details of what is happening right now. Time won't fly quite as much, but you stand a much better chance of remembering the fun.

As for whether any of this will stop the time from flashing by as we get older, Wearden's latest study has cast doubt on the idea that this is even true. In the study were two groups of people — one of people in their late 60s and early 70s, and the other young adult students. As with Wearden's other studies, people of different ages were interrupted with a smartphone app to ask them what they were doing, how they were feeling, and how quickly time was passing. They found that there were no significant differences in how young and old people perceive time in the moment. Whether you are young or old, the only thing that changes your perception of time in the moment is what you are doing and how you feel about it.

In short, it doesn't matter what age you are — if you are happy right now and/or are engaged in what you are doing, you report that time is going faster; if you are sad or bored, you report time is going more slowly. Emotion and attention is what matters for time in the moment. Age doesn't matter a bit. 'The idea that time goes more quickly in old people is not borne out,' says Wearden. 'I'm not sure why people say that, but they do ... Is it because it's not real, or is it because it is real and we're not clever enough to measure it?'

I think the reason people say that is because, like me, most adults feel like the months are passing by faster every year. This has nothing to do with the feeling of how time is in the moment — and anyway, it's impossible to measure, because there is no way to spool time back in to check how time felt 15 years ago compared with how it feels now. Research done about ten years ago by Wittmann, and then repeated by others all around the world, has found that the only consistent perception-of-time illusion is that as people get older they feel like the past decade went particularly quickly. Anything less than ten years, and time feels like it is passing at the speed that the clock says. And that, it would seem, is because it is.

To me, the take-home message from all of this isn't that we should fill our lives with new experiences to stop the last decade from seeming to fly by (it's not even clear if that is possible). Instead, it seems much more sensible to stop putting so much emphasis on the decades flashing by — because they aren't, even if they feel like they are — and to focus on the here and now instead. Easier said than done, definitely, but it is almost certainly worth the effort.

One final note of caution, though: in a recent study, Sylvie Droit-Volet looked again at what happened to people's perception of time as she manipulated their emotions with happy and sad films. This time, though, she split the group into two and told half of them all about the research linking changing emotions to passage-of-time estimates. And the results were: that the effect of emotion on time only works if you don't know about it. If you are aware that comedy is supposed to feel short, and horror long, then they won't actually feel like that. Perhaps the reason that my own results in table 3, on page 180, were so inconclusive was because, as Zakay pointed out at the start, I knew too much about what was supposed to happen. It is also possible that if you have read the whole of this chapter, then I might well have ruined the illusion for you. Sorry.

PART THREE

# A LOGICAL MIND

CHAPTER 6

# Number Sense Lost

*'Do not worry about your difficulties in Mathematics. I can assure you mine are still greater.'*

— ALBERT EINSTEIN

After all of that pondering on the nature of time and consciousness I'm ready to get my teeth into something a bit more solid. I'd prefer it if that solid thing wasn't maths, but there we are. For my last challenge, I have set myself the task of finding out if it is possible to improve on what I have long assumed to be a basic part of my brain's make up: I just don't have a 'numbers' kind of brain. If there is any basic function of my brain that is going to be tricky to override, I think this might be it.

It's almost certainly going to be worth the effort, though. Mathematical ability is linked to the capacity for logic and reasoning — and the mental gymnastics needed to think about abstract things like geometry might just feed into my burgeoning sense of physical space. And anyway, being rubbish at maths has proven to be quite embarrassing in my line of work, not least when I was an editor at *New Scientist* a few years back.

The tagline of *New Scientist* magazine is 'For people who ask why'. And that pretty much describes everyone who works there. These are the people who as kids would annoy their teachers by asking questions about everything in between mucking about. The sub-editors are no exception, plus they have an almost pathological need for stuff to be grammatically, literally, and factually perfect. Their job is to point out anything that is ambiguous, wrong, or deeply implausible, as a last line of defence before the magazine goes to press. When I was a newbie freelancer, I was terrified of them: I'd imagine them as a pack of hyenas, cackling over my beautifully crafted words and picking out all the juicy bits. In reality, they aren't like that at all; they're generally lovely — especially the late John Liebmann, chief sub at the time I worked there. He was so keen on getting things right that he would stop himself mid-sentence to correct his own facts or grammar. It would take ages for him to get his point across, which could be frustrating when you were battling a deadline, but when he did finish a sentence, you could be damn sure he was right.

It was during a conversation with one of the other sub-editors that number skills got added onto my wishlist of brain improvements. I was working on a graph to go with an article I was editing, and Sean, the sub, had come by to question my figures. My response didn't go down terribly well. 'Errr, okay. Let's have a look. I'm a bit rubbish at maths ...' Sean looked at me incredulously. 'What? And what are you doing about that?' he asked. I was a bit taken aback. 'Um, nothing really,' I mumbled. 'I just don't have a numbers kind of brain.' He stared at me for a moment and then finally shook his head. 'I just don't get how you can know that and not do anything about it.'

It was the first time that I'd been challenged on my assertion that I just don't have a head for numbers. I'd always just assumed that you either 'get' numbers or you don't. In my mind, I don't have the kind of mind that computes numbers, and, as a writer and editor, I was pretty much okay with that. You can't be good at everything — and I could get there in the end with enough time and a calculator. And I always asked someone to check my sums for me afterwards.

At the risk of sounding defensive, it's not as if I'm the only one that thinks this way. By one estimate, around a quarter of people have such an aversion to maths that they give up even trying, and panic in situations where they have to do maths under pressure, like when trying to work out a tip when the waiter is standing right there. At its worst it can turn into something that researchers call 'maths anxiety', which sounds like a proper diagnosis but isn't; it's just a more scientific-sounding description of the 'aaargh, I can't do this' feeling so many people get when given a numerical problem to solve. And this definitely sounds like me. I can almost feel the mental shutters come down when I see any kind of sum. Most of the time, I don't even try. I just reach for my calculator, even for the simplest arithmetic.

It wasn't always like this. When I was an 11-year-old, maths was embarrassing for a different reason. My maths teacher, Mr Griffiths, would ask a question and stare intently at the class, fingertips on the bridge of his nose, as he strode around the classroom, waiting for an answer. Eventually, when the silence got too much for him, he'd come to me for the answer. 'Come on then, Caroline, put them out of their misery.' I'd usually know the answer, although sometimes I'd pretend I didn't so as not to seem big-headed. I was pretty good at maths back then.

Soon afterwards, though, my number sense seemed to float away. I don't know how or why. All I know was that I stopped being top of the class in maths and started slipping a good way down towards the bottom. Ever since, I've pretty much opted out of the whole thing, and by and large I've gotten away with it. But since maths underpins most if not all of the science I report on, and comes up more in everyday life than I would like, I would really like to be good at it again.

Perhaps surprisingly, a basic understanding of mathematics comes as part of the package for humans, along with most other animals, from monkeys to rats, dogs, and even some fish. The fact that at the very least, most creatures can tell the difference between 'many' and 'not so many' suggests that it must be a key survival skill favoured by evolution. We humans have the added benefit of being able to manipulate numbers in the abstract to turn vague ideas about numbers into actual quantities. French neuroscientist Stanislas Dehaene has found that the less accurate, 'there or thereabouts' bits of maths processing uses visual and spatial brain areas, whereas the exact stuff requires the same areas as language processing. So to some extent, being a words person and a numbers person are kind of the same thing. No excuses there, then.

Oxford maths professor Marcus du Sautoy also argues that there is no such thing as a non-maths brain. Even if you count people with dyscalculia, the maths equivalent of dyslexia, which affects around 5 per cent of the population, we are all mathematicians, he argues, because mathematics is basically the ability to spot patterns in the world. You might not be good at arithmetic, in other words, but pattern-spotting is such a general skill that there is a lot about maths that you can master. Pattern-spotting is such a key survival skill that it has been selected for time and again by evolution, says du Sautoy. 'If you saw something symmetrical, it was likely to be the face of an animal, and either you could eat it or it could eat you. Either way, those sensitive to symmetry survived,' he wrote recently in the *Guardian* newspaper, in an article that pooh-poohed the whole idea of there being such a thing as a non-numbers brain.[1] 'Similarly, humans with a good sense of numbers could tell whether their tribe was outnumbered or not, which would inform the decision to fight or flee.'

Clearly, though, people do vary in their maths ability. We are not all Oxford professors of mathematics, after all. The question of why we differ so much is something that interests Roi Cohen Kadosh, a cognitive neuroscientist also at Oxford University. He studies how these hard-wired tendencies get shaped by learning, and also how factors like personality, reasoning skill, and attention affect mathematical ability. After hearing my experiences, he suggests that loss of confidence is probably the main problem — as I have heard a few times now, freaking out about not feeling capable of doing something is a sure-fire way to rob the brain of the resources it needs to do it. Could it really be that my conviction that I am terrible with numbers is the reason that I am? And could that be the reason why a recent report by the Organisation for Economic Cooperation and Development (OECD) found that 54 per cent of boys and 65 per cent of girls consider maths to be stressful.[2]

If that's the case, then perhaps a bad head for numbers will be easier to turn around than I thought, and might not even require too much messing with my brain at all.

Roi puts me onto one of his students, Amar Sarkar, who recently did a study into how brain stimulation might help people to get over their maths aversion enough to unleash their hidden maths genius.[3] Amar wouldn't put it like that, though. When we meet at Oxford University, he is cautious and reserved, and he speaks slowly and deliberately to make it impossible to draw any overhyped conclusions from his research so far. He is particularly keen to underline the fact that what I am doing — trying things out on myself — is not science. 'It will be scientifically invalid, but still, an interesting experience for you,' he says. Amar is young and just beginning his scientific career and hasn't even run the gauntlet of being misquoted many times in the science media, so I'm impressed that he seems to inherently understand that, given the wrong information, a journalist might well claim that he is unleashing people's hidden maths genius. So I probably shouldn't. But it isn't actually too much of an exaggeration. In a recent study where Amar compared two groups of people — one group with an emotional aversion to maths, and the other without — he found that, while the low maths-angst group were a bit better at maths in real life, the high maths-angsters were also pretty good: well above average, as you might expect from Oxford University students, who made up most of the study volunteers. The reason they struggled at maths seemed to have very little to do with their actual ability.

Nevertheless, when asked to work out whether a simple sum (for example, 8+2 = 10) was true or false, people who were scared of maths were significantly slower. And when they measured levels of the stress hormone cortisol, the maths haters had significantly higher levels than their more confident counterparts. But — and this is where it gets interesting — when they boosted the electrical signals in the right prefrontal cortex with electrical stimulation (this is above the eyes, at the top of the forehead, and involved in control of emotional reactions), they found that it not only reduced their cortisol levels significantly but also allowed them to react about 50 milliseconds faster to the maths questions. As Lila Chrysikou told me in Kansas, 50 milliseconds is quite a lot to a psychologist — but in terms of using this in real life, the cortisol changes are far more impressive. Lower cortisol means feeling less stressed out, which is something that you could definitely feel. Might this simple intervention, shutting down stress a little bit, really help people love maths? Or at least not hate it quite so much?

Here's where I find out. Having tested me out and concluded

that I fit into the category of a maths-anxious person, Amar offers to give me a week's worth of stimulation and a bit of cognitive training, just for fun, to see if it changes anything in not only my scores but also how I feel about maths.

I have been stimulated in the brain lots of times now, and it never stops making me feel a bit nervous, especially when Amar tells me that he is going to use a more powerful and long-lasting version called tRNS (transcranial random noise stimulation).

I'm cheered up, though, by the fact that Amar is strapping the electrodes to my head with a lovely bright-blue-towelling sweatband. I comment that it's going to be a fabulous Dire Straits look, then wonder if this will make any sense to Amar, seeing that he was born four years after the classic *Money for Nothing* video came out; plus he grew up in India, where I have no idea if Dire Straits were even a thing. But he grins broadly and tells me that, although Dire Straits weren't huge in India when he was growing up, his parents are big fans of stadium rock, and he was raised on the likes of Bruce Springsteen and Dire Straits. I briefly glimpse a less-reserved side of Amar as we chat about the joys of dad music, and ponder what Mark Knopfler is up to nowadays.

Then it's onto the familiar routine of baseline tests (pages of basic arithmetic questions that get harder and harder, and I'm not allowed to skip any), plus various measures of my working-memory capacity.

Then he straps me into the brain stimulator and turns on the current. 'Do you feel enhanced?' he asks. 'No,' I reply. 'Should I?' 'No,' he says, mysteriously, and then I start the test.

Unlike when I was in Kansas, I don't feel any kind of wonkiness, buzz, or other changes in myself. That might be because when Lila plugged me in, she was turning down activity in the prefrontal cortex, whereas Amar is turning it up. Perhaps it's easier to notice when you lose brainpower, but a bit extra doesn't necessarily seem that different. I certainly don't feel like any kind of genius, that's for sure. Even so, I quickly settle into the test and, once it becomes clear that I can manage the level of mathematics required to tell whether a sum is true or false (for example, $9 \times 3 = 27$), I relax and start racing through them. It seems to help if I say the sums out loud, so I start muttering to myself whenever Amar leaves the room.

I do the same for the next few days — and on days two and three, Amar gets me to do some cognitive training that, in an experiment he has just finished and which isn't published yet, has also had some effect on maths ability. It's quite fun — I have to pretend to be a worker in a robot factory, and have to make various on-the-spot decisions about what to do with robots that come along the conveyor belt: if it has broken arms, press left; if it is red, press right. If it has a yellow light around it, press nothing. I recognise it as taxing working memory plus some of the mental-control skills I lacked back in the early days in Boston, and am struck by how much easier I find it to make these kinds of fast mental decisions.

In the Betty test, I found it impossible to change my mind about pressing a button once my hand had started moving. Now it's not a problem. Perhaps this is no coincidence: pretty much everything I have done since has been relevant to the prefrontal executive control bits of the brain. Coping with the robot task might be proof that I have gained a small amount of control. Or it might be that I am getting a boost in this area from the stimulation, which I am today using via a rather fetching rubber cap. I later learn that the theory behind this is that training working memory transfers to a skill that needs working memory (maths). It's back to working memory again, albeit in a slightly more interesting game format. It's perhaps no surprise. I mention to Amar that executive functions keep cropping up. 'Most of your book is executive functions,' he confirms.

After that, I have a whole afternoon to kill in Oxford, so I head to one of the city's many bookshops to look for a maths revision guide. Amar isn't keen on this idea because it adds another factor to the experiment that he didn't plan for, but concedes that since this isn't a proper study, and given that I'm only having two robot-factory training sessions — rather than several weeks, as they would in the real study — adding a few minutes of maths practice here and there probably won't make much difference. Amar tells me that sometimes you don't see transfer for several weeks after training anyway, as it takes time for the changes to make themselves apparent.

I had thought of picking up a revision guide a few weeks earlier at home, but didn't want to ruin my baseline scores by practising too early. I had also been thwarted by the full force of my aversion to maths. I walked into my local bookshop, headed to the revision guide section, and pulled out a maths book aimed at teenagers. I opened it to a random page and ... well, the pages might as well have contained images of rotting corpses. I physically recoiled, shoved the book back on the shelf, and found myself walking out of the shop before I realised I had even told my legs to move.

This time, in Oxford, I decide to ease myself in more gently, and so pick an exam revision guide aimed at 10-year-olds. That night, on the train to see a friend in nearby Didcot, I have a go, one question at a time, taking it slowly, and being sure to check the answers on the cheat sheet as I go along. It could be that having been zapped earlier that day did something to my brain, but weirdly I find myself actually enjoying it. And every time I get the answer right, my confidence grows. My final score is 96 per cent. Not bad at all.

Sure enough, a few weeks later, when Amar sends me my results, they seem to suggest that my skills have indeed improved. On the baseline maths tests, where I had to wade through pages of multiplications, long divisions, and the like, my baseline score was

98. After the stimulation and training, it jumped to 106. It doesn't sound like a big improvement, but Amar seems impressed. 'It's an 8.1 per cent improvement. For just two sessions, that seems pretty sizeable.' Based on other data, he tells me that a rough estimate for the practice-effect is around 2 per cent.

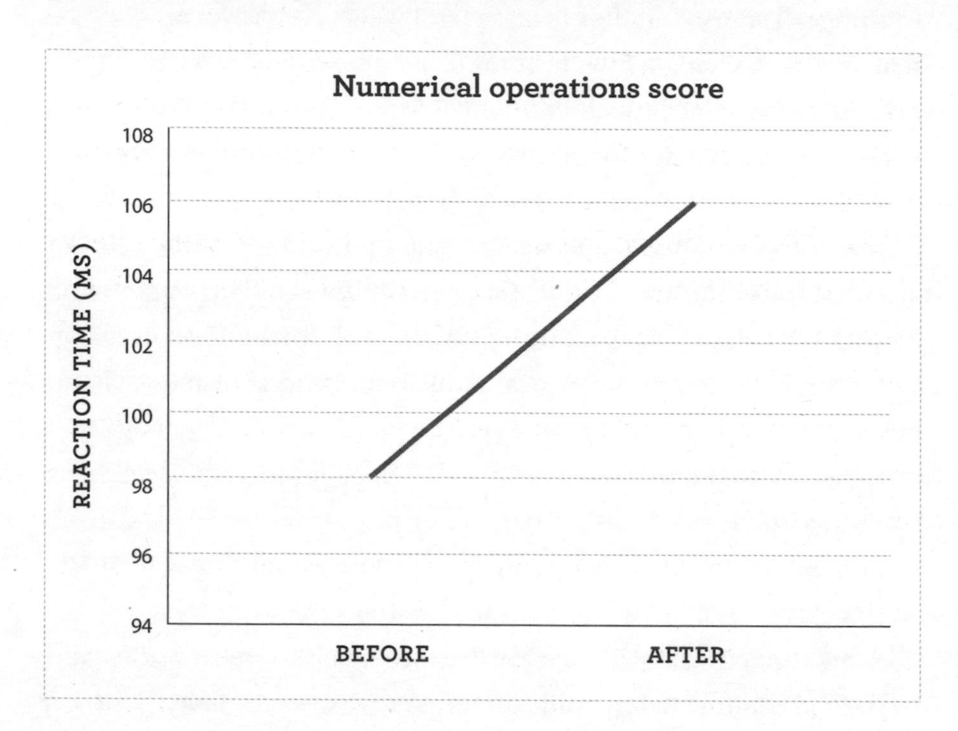


Fig 17. Numerical operations score.

He is, however, keen to point out that this doesn't necessarily mean that the other 6 per cent came from the intervention. 'This is a sample size of one, and there could have been any number of reasons that you would *want* to improve in the second session,' he says. Expectation or regression to the mean (a strange law of statistics that says that a second score on anything will be closer to average than the first, whether you do anything to change it or not) could have played a part.

One thing I hadn't paid a huge amount of attention to at the time was that, before each sum appeared on the screen, a word with either negative or positive connotations (a 'prime') flashed up briefly in front of me. In a previous study that Amar's is based on,[4] people with high maths anxiety were quicker to respond when the prime was a negative word like 'useless' or 'failure'. This result was a surprise: usually, having a boost of positive thinking helps people to do their best work. Instead, with maths anxiety, it seemed to resonate with how the people saw their own abilities, which somehow made them perform better when the prime was negative.

When Amar repeated this study, he didn't find the same effect in his sample (he says that it might be because his sample was mixed sex while the original was women only. Women are known to be more likely to be maths anxious than men). For this woman, though, the negative words didn't make any difference: my reaction time was more or less the same whether the word was a compliment or an insult.

The fact that negative primes do affect some people suggests that they are working as another kind of unconscious bias. In chapter two, I found that my focus was being tugged towards disapproving faces, while skipping over the happy ones. Practising doing the opposite, using online training, seems to have done the trick in re-writing that bias. The ultimate aim for maths-anxiety training is to do something similar with the way people feel about their maths ability. 'This will be the really interesting thing ... Can we cause people with high maths anxiety to stop benefiting from the negative prime and start benefiting from the positive prime?' It hasn't been done yet, but that's definitely the aim of this kind of research — to make people maybe not love maths, but at least not have a negative emotional reaction to numbers, whether that be conscious ('I don't do maths') or unconscious ('Oh look, I seem to be walking away from the maths section of the bookshop rather quickly'). Amar agrees. 'Yes. I guess that would be the ideal outcome,' he says, cautious as ever.

While I didn't show any sign of being affected by the positive or negative words that came before the sums, during stimulation, the speed at which I decided if the sum was true or false increased by 200 milliseconds compared to baseline. Again, this is more impressive than it sounds. 'A 200-millisecond improvement at no cost to accuracy in performance is considered huge. In comparison, the improvement in my maths-anxiety paper was just about 50 milliseconds,' says Amar.

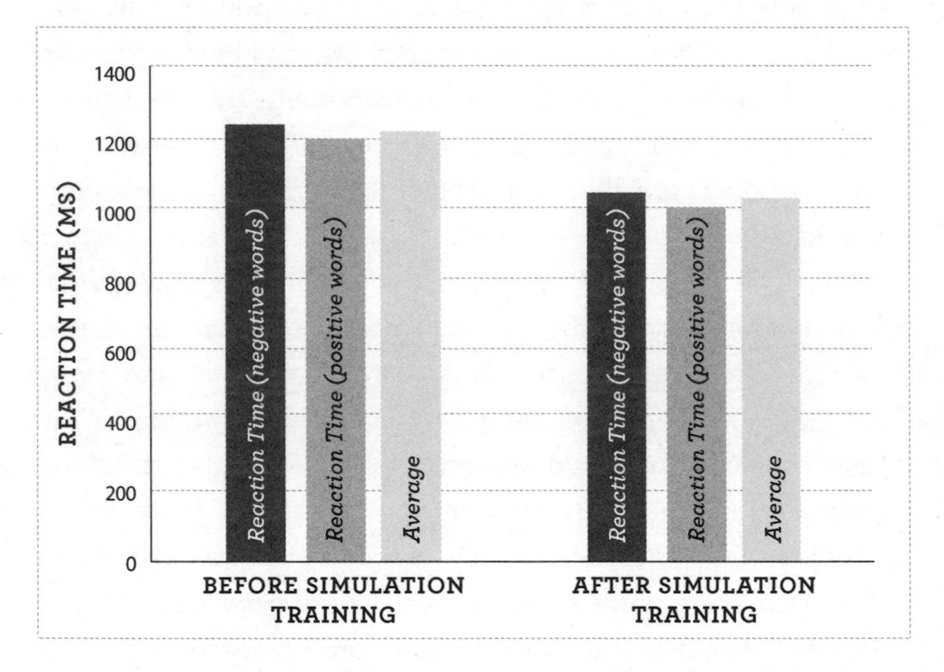


**Fig 18. Reaction time in milliseconds before and after stimulation and training. My accuracy score was the same before and after (93.5 per cent versus 93 per cent), suggesting that I didn't speed up at the cost of accuracy.**

Again, there are a whole heap of caveats to add here. To quote Amar when he sent me my results: 'The main thing is that performance on the second session was much better than the first. Of course, this doesn't mean that it was due exclusively to stimulation. To truly determine whether it was the stimulation, we would need about 120 participants, 60 of whom complete the sessions with real stimulation, and 60 of whom complete the same thing with placebo/sham. If the improvement is greater following real compared to sham, then we can say that stimulation produced the effect.'

We didn't measure my cortisol levels and so don't know if I, like the volunteers in Amar's study, would show a lowered stress response after being stimulated. I certainly felt less terrified about the sums in the 'after' testing, but that could just as easily have been because I was more familiar with them and knew that I could handle some if not most of them. And getting 96 per cent in the revision guide didn't harm my confidence either.

In the end, though, it doesn't matter whether it was the stimulation or practising maths that made me feel less terrified of the test. As Amar points out, doing maths puzzles is a form of brain stimulation. In other words, you don't need to wire your brain into a battery, but it seems to help.

Whatever starts it off, once that confidence starts to build, it sets up a virtuous circle. And if Amar is anything to go by, it is absolutely possible to get better at maths with a bit of effort. On day two of testing, as we head to the testing room, he admits that, until recently, he had a pretty bad phobia of mathematics — something that is not terribly useful if you want a career in science. Thankfully, he tells me, it's not a problem anymore. 'What did you do?' I asked, hoping for a pearl of wisdom that will inform what I do from now on. He grins at me as the lift doors open, and raises his eyebrows. 'I practised.'

Practice, then. That's about as high tech as it needs to be as far as maths is concerned. Sean-the-sub was right all along: I just need to stop being fatalistic about having a 'non-maths brain' and do something about it. It harks back to the idea of a growth mindset, which I encountered in the debate about working-memory training. Whether people benefit from cognitive training or not has a huge effect on whether they think it is possible to improve, whatever their current ability. On the other hand, if you don't think you can do anything about a skill, you probably won't bother to work on it, and your predictions about being bad at it will come true.

Nevertheless, the stimulation did seem to have an effect, and I'm intrigued to find out what exactly it might be doing to help things along. After my trip to Oxford, I call Roi Cohen Kadosh and ask him. 'It's a really good question,' he says, suggesting that no one else knows either. 'I can say what we *think* ...

'There are some results showing that we can modify neurochemicals in the brain — some of them are associated with neuroplasticity,' he tells me. 'Then we are affecting connectivity between different brain regions with stimulation as well [and] affecting the consumption of oxygen and metabolites.' Is this all happening at once, or does one thing happen first and set off a domino effect of brain change? 'It's hard to know,' he says. 'Maybe it's all of them working at the same time.'

Then there is the possibility that stimulation gets the brainwaves going in a particular frequency that allows for concentrated thought. Setting the stimulator to the gamma frequency of 40Hz, for example, might encourage natural brainwaves to follow suit. Gamma waves are at work when we are concentrating hard and pouring all our mental energies into a taxing mental puzzle.

Whatever is happening exactly, the basic theory is that it makes the region under the electrode more active than it was before. Given that the region being targeted in this experiment — the dorso-lateral prefrontal cortex — is involved in regulating negative emotions, it could be helping to do the job more effectively. This would explain why stimulating the prefrontal cortex helps maths-anxious types. If they are slower because they are processing an emotional reaction at the same time as trying to work out whether 8 + 6 = 12, a bit of extra brainpower is bound to help — adding more juice just helps the whole process along.

In this case, the key thing is how well you can dampen down the 'aaaargh' reaction to get your thinking brain into gear. If you can get rid of that obstacle, it frees up some mental capacity that you can then use to do sums. Unlike most of the hype about brain training and stimulation, it isn't always about adding capacity — it's about releasing the capacity that you already have, or removing a block that has no business being there.

Brain stimulation, then, might not be strictly necessary, but new evidence suggests that it does seem to make a difference. In a recent study, Cohen Kadosh's team (this time not including Amar) found that doing mathematical training improved scores in all of their healthy volunteers, but adding tDCS gave one group a significant boost over the other, who received sham stimulation instead.[5]

Having had a fair amount of brain stimulation now, I can kind of see why braver people than me think that this is something they could use at home. When you see a scientist dip a sponge into salt water, measure roughly where the right bit of brain is inside your head, pop a towelling headband on top, and then turn on the current, it doesn't seem out of the realms of possibility to do the same thing at home. It's certainly easy enough to get hold of a tDCS machine, if you have access to the internet and a couple of hundred quid. So what's the problem with it?

Well, for a start, Amar's study suggests that sometimes stimulation can make things worse. In his experiment, only the people who hate maths got any benefit from stimulation, in terms of their reaction time and cortisol level. People who were confident about their ability got worse after a zap to the prefrontal cortex: they slowed down, and were less able to keep their cortisol levels at a nice, low, unstressed-by-maths kind of level. Both groups, the confident and angst-ridden volunteers, performed worse at standardised tests of attentional control. So you can easily strap a battery to your head, but you might just make yourself slower and more stressed, not lightning fast and confident. 'It's not a free lunch,' says Amar. 'If you enhance one process, it might come at the cost of another.'

Because most studies are done on large groups of people, and the data all put together, these subtle differences often go unnoticed, says Roi. As a result it has barely been raised as an issue, in science or in the burgeoning home-tDCS market.

It is also possible that the bit of the brain that needs stimulation will change as the learning process happens — something that scientists haven't even gotten a handle on yet, let alone home users. Roi and his team are looking at whether stimulating the prefrontal cortex is good at early stages of learning — it helps with the control needed to make a deliberate effort to learn, but when a person gets better at the task, and is mostly retrieving information from memory, areas further back in the parietal lobes might benefit from a boost instead.[6]

More importantly, Roi says, there is no evidence that using tDCS, or any other form of brain stimulation, every day for months on end is good for you, or even safe. To know that for sure would involve stimulating a group of people every day for months and keeping a very close eye on them.

'I would not run this study,' he says with a wry laugh. 'I would not feel good about stimulating my own brain for three months. I am not going to do that for anyone else!'

All in all, there are too many reasons to be cautious, he concludes. 'We don't know who it fits, who is going to benefit from it the most; we don't know if it is safe to use it for an extended amount of time. Stimulation by itself doesn't seem to be effective — you need to combine it with cognitive training if you want to induce long-term changes rather than just improving on the task you are doing at the moment. All of these together doesn't make me say "Hey, you know what? Let's go and do this!"'

I trust his judgement on this — particularly as he says he has been offered lucrative opportunities to put his name to home tDCS, but has so far resisted the temptation. 'I think that this is not, at the moment, the right approach. We need to know more.'

Instead, Roi lobbies for the regulation of commercially available brain-zapping headsets, which currently don't come under legislation for medical devices. In the same way that dietary supplements don't come under drugs regulations as long as the manufacturers don't make any specific medical claims for them, home-tDCS kits are outside the current regulation. Which means that there is an experiment going on in heads all over the world, but one that is not being safety-checked or monitored in any way.

And, as a final point on brain stimulation — which also relates to everything I have done to change my brain — there is this basic fact: no one gets to be superhuman. It may be fun for the likes of me to think, 'Great, we have this neuroplasticity thing, let's use it', but the evidence for 'cognitive enhancement' is more about ironing out individual differences stemming from genetic differences or what we have learned along the way. It is not going to turn any one of us into a super-computer (or as it says in one of Roi's papers, it's not going to take us 'beyond the species-typical level normal range of functioning'[7]). In the same way that no matter how much protein I ate as a child, I was never likely to grow to more than five feet tall (the genetic limit for women in my family, it seems), the brain is plastic, but there are limits.

Taking all of that on board, has it been worth the effort of putting myself through these extra maths classes when I have a perfectly good calculator on my phone? For me, yes. It has been a huge confidence boost to find out that I have plenty of capacity there, waiting to be used, if only I can get past the initial aversion and start to enjoy the puzzle.

Which brings me to logical reasoning — a skill related to, but not the same as, mathematical reasoning. Like maths, and indeed nearly everything I've been working on, this is not something that comes particularly easily to the human brain. It relies on executive functions (particularly working memory) to keep a deliberate train of thought going long enough to come to a logical answer.

As far as maths goes, it was a surprise and relief to find that I didn't exactly need to change my brain at all — I just needed to realise it was in there somewhere, buried under a layer of brain-sapping low confidence. I am now wondering if this lets me off the hook about improving my logical reasoning skills, too — not only because if my maths skills are fine then my reasoning skills probably are, too.

There is no denying that I have always been something of an emotional person (see chapter two). This basic feature of my personality often brought me into conflict with my stepdad when I was a kid because, to him, logic is everything. I've lost count of the number of times he told me that there's absolutely no need to get all het-up and emotional about stuff — just think it through slowly and logically. If the answer to your problem makes logical sense, then it must be right. It might have something to do with him being brought up by a sergeant-major in the military police. Or maybe he's part Vulcan. Either way, it's fair to say that we don't entirely see eye to eye on the role of emotions versus logic in deciding how to live.

Actually, there is some pretty good evidence that I am at least partly right about the importance of emotions. A particular region of the brain's prefrontal cortex seems to be important for combining emotional information with more logical information, when we make decisions. Studies of people with damage to this part of the brain have shown that, when they are given a choice between two options, and there is no logical advantage to either of them, they struggle to choose at all. When there is no logical reason to choose, it seems, we rely entirely on our emotions to guide us. Without our illogical gut feelings and emotions, we would be lost.[8]

On the other hand, studies have shown that, when it comes to making rational decisions about the people or issues we love, emotions are no help whatsoever, often leading us to conclusions that are just plain bad for us. I'm reminded of the experiments I did with Amar, in Oxford, where my terror of maths left me unable to think about the problem in hand. Allowing your emotions to get in the way can be disastrous for mathematical thinking, and probably logical thinking, too. So, it seems that there is no easy answer to the 'logic versus emotions' question. Ideally, we all need a bit of both.

Then there are the unconscious biases, which influence even our most logical decisions, whether we like it or not. Researchers at Harvard University's Project Implicit are trying to measure these biases via a series of online tests on their website: projectimplicit.com. These short tests provide you with immediate results on how your unconscious biases might be colouring your decision-making. In my experience, they are quite revealing.

I have always thought myself to be immune to the idea that high-fat foods are inherently bad — logically, I know that a bit of everything is fine in moderation, and I try to make my food choices accordingly. I can't see anything wrong with eating chips or chocolate if that's what I fancy. According to the Project Implicit test about attitudes to foods, though, I strongly associate words like 'shameful', 'disgusting', and 'unacceptable' with high-fat foods like cakes and biscuits, and positive words like 'healthy' and 'success' with low-fat foods like fruit and vegetables. And while, consciously, I am equally unimpressed by all organised religions, according to the Project Implicit test on feelings about religion, I feel most positive about Buddhism; then Christianity; then way further down the scale are Judaism and Islam. This makes me wonder how much this is feeding into my daily life without me being consciously aware of it. Am I a closet Islamophobe? Have all the headlines gotten to me despite my lefty liberal leanings? And has a combination of my Christian schooling and recent love affair with yoga and meditation pushed Christianity and Buddhism to the top of the list?

Not all of my implicit assumptions are so much of a shock, though. I am pleased to find that I associate women and men equally with the home and the workplace, and have no bias towards women in science (this might be because I talk to a lot of psychology researchers, where women are better represented than in other areas of science). And when I took the 'attitudes to homosexuality' test I was slightly more likely to associate words like 'joyful' and 'lovely' to gay people than to straight people.

How much these implicit biases impact on day-to-day thought processes isn't clear, but research does suggest that, when our beliefs contradict the evidence before us, our minds do backflips to try to make our beliefs stand up. This is why people are susceptible to conspiracy theories even when the evidence is right in front of them. The only way to align the two is to get them out in the open and take a long hard look at the assumptions you don't know you are making. I recommend giving it a go — it's fascinating.

'Know thyself' is a theme that has come up time and again as I have been trying to improve my brain skills. If you know what your limits are and why, then you stand a much better chance at tackling any problems at the source.

And so I decided not to pursue logic training. When I started this project, I thought that my emotional circuits were quite strong enough and that what I needed to do was to strengthen the logical circuits to balance them out. Now, though, I'm not so sure. I keep remembering what John Kounios, the creativity researcher I met in Philadelphia, told me about creativity. His research has divided people into two broad groups depending on the way they tend to solve problems. Some people fall into the category of 'analytical' problem solvers, who tackle a problem by slowly and steadily working through the options. These people, Kounios found, have relatively more activity in their left hemisphere (it seems there is a smidgen of truth in the 'left brain = logic' idea, even if there is way more to it than that). Other people, who Kounios calls 'insightfuls', tend to solve problems by insight and have relatively more right-hemisphere activity at rest. This method of problem solving is a bit hit-and-miss because they tend to have nothing to show for their efforts until a solution pops into their head in an 'aha' moment. According to Kounios, there is some tentative evidence that these tendencies might be inherited and are quite stable over time, suggesting that we come set up to either be logical or creative types and tend to stay that way throughout life.

In Kansas, I did lots of tests of creativity, and several insight tests, with Lila Chrysikou. Without wanting to blow my own trumpet too much, I came out way above average on all of them. We didn't measure which side of my brain had more activity at rest, but all the other results point towards me being more of an insightful type than a logical Vulcan.

Do I really want to push myself away from that and towards a more logical mind? That seems like a step too far away from the person that I naturally am. There have been parts of my natural state that I have been happy to change (anxiety and distractibility being two of the most obvious), but even though it's probably possible to override my creative bent to get more logical, I'm not sure I want to; I quite like being a writer ...

For anyone wanting to boost their logic skills, though, there is evidence that if you spend a lot of time practising logic problems (and get the answers right), it will have a physical effect on the structure of your brain. In studies at the University of California in Berkeley, when prospective law students took a crash course in logical reasoning as part of their preparation for their Law School Admission Test (LSAT), their brains gained connections between the frontal and parietal lobes — two regions thought to be important for logical trains of thought — after just ten weeks.

The volunteers were set logic puzzles every day for ten weeks, which is quite an investment of time in something you are not even sure you want to do — and anyway, when I asked her about it, the lead researcher in this study, Silvia Bunge, wasn't able to have me in her lab to do the experiments anyway. So I can't offer any specific advice on whether this works, but anyone who wants to take the challenge can find LSAT revision papers freely available online.

Instead, I found some online tests of logical reasoning of the type that prospective employers use to torture well-meaning graduates, and gave them a go.[9] My score came out as middle of the road, average — fine, but not stellar. And you know what? I'm okay with that.

Interestingly, when talking to a friend about my decision to leave my powers of logic well alone, I realised something. If I'm choosing not to train my brain in a given direction for fear of it working too well, it must mean that I am starting to believe that brain override is possible. I started off imagining that I'd find at least some things that couldn't be changed, no matter what you do, but I have to admit that the brain's adaptability has come through every time. Some of the changes I have made came down to simply changing my mindset, others to practice, some to embracing what is already there, and some to tackling unconscious biases that I wasn't previously aware of. But for all the different routes to changing my mental workings, one thing seems to be coming through loud and clear: work on what you want to change, and after a few weeks things will start to happen.

### *OVERRIDE YOUR RUSTY MATHS BRAIN IN FOUR EASY STEPS*

1. Do some easy maths — use pencil and paper; take your time
2. Do some slightly harder maths — on paper; take your time
3. Realise you can do it if you put your mind to it
4. Repeat steps 1 to 3 as necessary

# PART FOUR
# WHAT NOW?

CHAPTER 7

# My Brain on Override

*'Override: to interrupt the action of (an automatic device), typically in order to take manual control.'*

— GOOGLE DEFINITION

It's time to tackle the million-dollar question: do I think I have changed my brain? Or — given that everything changes your brain and that none of us are sitting here today with the same brain we had a year ago — have I made any useful, deliberate changes that have moved it towards the kind of brain and mind that I was aiming for at the start of this experiment?

Yes, I think I have.

But before I get to what I think has changed, how I plan to keep it going, and how anyone could do something similar, I want to raise the possibility that we have been thinking about brain change all wrong. The reason I say this is because the way that I think my brain has changed is not at all what I was expecting.

When I started this project I, like most people outside of working neuroscience labs, thought that what I needed to do was to tone up a few brain areas and improve on a few key circuits in my brain — a mechanistic approach, if you like, that thinks of the kilo or so of mush in my head as an engine that needs a bit of work: upgrade a component here, insulate a wire there, and get the whole thing running more smoothly.

When I tried to turn this idea into practical action, though, it became obvious that this view of the brain is missing an important part of the picture. While there are definitely areas of the brain that specialise in certain skills, anything worth doing requires the activity of many, many different brain areas. With that in mind, what is most important is not activity in certain areas, or even the thickness of the wiring between them, but the way that the brain puts these different streams of activity together to create something that is greater than the sum of its parts.

Taking this view of the brain into account, it doesn't make sense to try and strengthen particular areas or even the circuits that they are part of. A far more useful thing to do would be to work on flexibly engaging different circuits of the brain based on which is right for the job right now, with the aim of shifting between the best state of mind for the job at hand.

Here's one example from my recent endeavours. The same brain areas are involved in both sustained focus and in creativity, but they are activated in a different pattern. Sustaining focus involves keeping the frontal bits online to concentrate on the job, and allowing the mind to wander occasionally — but not too far. Creativity involves unhooking the frontal control parts for a bit, to let the mind run free, while remembering to bring the focused mind back now and again to check that your ideas are not too bonkers. These two states of mind have a lot in common, but they are also very different, and having access to both of these states requires control over the switch between the two.

Other areas I have worked on brought similar challenges. Navigation requires the ability to switch from representing a cognitive map in memory, to creating a mental image of yourself in that map, facing in a particular direction. This is a complex bit of computation that involves a huge range of brain areas. Controlling stress and anxiety, too, comes down to being able to switch from a danger-spotting mode into a more relaxed and mindful state, to investigate the situation from a mental and emotional distance. In all of these areas, strength is great, but controlled flexibility is even better.

It has been slowly dawning on me throughout this project that when you think of it this way, getting override privileges for particular skills and behaviours has less to do with tinkering with particular parts of the machine and more about learning to drive it — if such a thing is possible. And, from my experience this past year or so, it certainly feels like it is.

This view of mind control owes a lot to a newer wave of neuroscience research that has been ticking away while the rest of us have been buying into the idea that our brains can be trained like a muscle. In recent years, the idea that a particular brain area, or even a group of brain areas, is in charge of complex mental abilities has become a bit old-hat — an updated version of phrenology, the 19th-century 'science' that mapped different mental faculties to specific 'organs' in the brain, which could be read via the size of bumps on a person's head.

As brain-imaging methods have become more sophisticated than head-bump-reading, though, it has become possible to look not just at the size of certain brain areas or brain activity within them but also at the fibres linking different parts of the brain to others, and the activity running along them. The ultimate plan is to use this information to map large parts of that wiring diagram — the connectome — in human brains, and ultimately to link differences in wiring and activity to the variation in behaviour and mental ability. We are nowhere near to this yet, but the focus has nevertheless begun to shift away from what different parts of the brain specialise in, and towards how activity in the networks that link them fluctuates over time.

Driving these larger-scale networks deliberately sounds as if it would be even more difficult than finding an exercise that taxes a specific brain muscle. To my mind, though, it actually makes the task easier. As I have been working through my shopping list of brain changes, a handful of linked-up brain areas and the mental states they lead to have kept cropping up. And when I succeeded in making changes, it almost always involved getting into the right 'state' for the job — or at least noticing when I'm not in it and doing something about it.

I'm wary of sticking my neck out too far here for fear of adding to the already bulging back catalogue of 'neurobollocks', but if I had to summarise the mental states that I found were the most useful to have control over, I would say they are as follows:

| Mental state | Good for ... |
|---|---|
| Relaxed and ready | Lasting focus, developing creative ideas, stepping out of time |
| Mindlessness | Hypofrontal creativity, and the gateway to other states |
| Mindfulness | Coping with stress and to slow time in the moment |
| Effortful thinking | Maths and logic problems, idea evaluation |
| Anxiety | Nothing. Bad for everything |

**Table 4. Mental states for override.**

In my own life, I have gotten into the habit of using this as a framework for understanding and controlling my mental state to get whatever needs to be done, done. If I wake up tired and am finding it difficult to concentrate, I'll either choose a creative, big-thinking kind of task that morning or, if there is focused work that can't wait, I'll go out for a brisk walk with a strong cup of tea and come back to my desk in half an hour. That approach seems to make me relaxed and ready in a way that sitting, staring at the screen, and fighting to focus never has. When friends ask me what my biggest lesson from this experiment has been, I tell them that my main conclusion is 'knowing when to say *fuck it* and go for a walk'. I'm joking when I say this, but only just.

Similarly, if I notice I'm distracted, I'll consciously direct my attention onto my thoughts and body to see if there is anything that needs my attention — perhaps an unhelpful undercurrent of fear of failure, or a tense jaw that isn't strictly necessary for thinking or reading. Then I'll give myself whatever is necessary — a break, a walk, a cup of tea and a biscuit, a pat on the back from reading something I have written before that came out well, anything that will ease off the mental pressure — before directing my attention back to the task at hand. If I'm feeling rushed, or as if the day is getting away from me, I'll stop and consciously direct my attention to the details of what is around me, until it slows down to normal speed and I feel more in control.

I'm not saying I've morphed into a calm and unflustered productivity machine, but I have to say that, nine times out of ten, thinking about my mental state in this way helps me to feel more in control — as if I am choosing my mental state rather than letting my default state rule me. A case in point: I am writing these words while only two weeks away from my deadline for this book, and I still have a frightening amount to do. But whenever anyone asks how it's going and whether I'm going to meet the deadline, I find myself calmly saying that there's a lot to do, but it's all good; I'm getting there. That's not something I would have said a year ago about a magazine article, let alone a book. When it comes to dealing with stress, something has definitely shifted.

Whether anything has changed physically in my brain to make this happen, or whether I am using the existing wiring slightly differently, is harder to say. Curious, I contact Martijn van den Heuvel, of Utrecht University, in the Netherlands, to see what he makes of all of this. He is one of the researchers leading the charge to describe how the brain's wiring is organised into networks, and how variations in the way that people's networks are set up affect brain function and behaviour. What he and his colleagues have discovered fairly recently is that the brain's wiring is organised around 12 highly connected regions, or hubs (six matching pairs, one on each side of the brain), which integrate information from all around the brain as it moves into and out of various mental states. A few networks are described as 'rich clubs' because they have more connections to each other than other regions of the brain have. I'm wondering if these hubs and rich club networks relate to the mental states I have been working on, so we connect on Skype for a chat.

I'm used to being the one asking questions, so it throws me slightly when Martijn starts by asking me whether I feel as if anything has changed in my brain. I put my theory to him that perhaps the size of specific brain regions and the wiring between them probably hasn't changed, but that I think I have learned to engage particular networks more efficiently. I suggest that maybe I haven't so much changed my brain as change the way I use what was there all along. I burble this out, bite my lip, Kermit-style, and wait for the verdict.

'I think I can deal with that ...' he says, nodding slowly. 'But as a biologist, I would argue that also some structural changes would have to be made. Perhaps not at the level we are used to seeing with

MRI, but if you would be able to zoom in to the individual neurons and the individual synapses, then there still needs to be some sort of chemical or anatomical background to this change.' It's what I was talking to Heidi Johansen-Berg about all those months ago. You wouldn't be able to see anything in a brain scanner, but if you believe that the brain can change — and there is good evidence that it can from all those animal studies — then something must have happened in there.

What, exactly, has happened is difficult to say, but it might have something to do with being able to integrate information from different brain areas more efficiently, and perhaps with changes in the way that different parts of the brain talk to each other.

'Many of these properties you mention are to do with a balance between different skills of the brain,' says Martijn. 'If we assume that we have different brain regions that all have different functions, there needs to be some kind of continuous negotiation between those skills.'

So if you imagine a few minutes in the life of a brain, you might start with a resting state, where the default-mode network is the most active, chugging away doing nothing very much. After a while, it will drop off and a local network will come to the fore — maybe you hear something, and the auditory processing parts of the brain become active. If that local network activates one of the hub regions — perhaps part of the salience network, because it is particularly relevant for important noise — the information flows to other hubs. Perhaps it will switch on the visual areas, for example, to look for what the noise means. All of this information feeds into an understanding of what is going on. Then this activity may subside and the default-mode network comes up again. 'It is like a consequence of how the brain is wired that you would have this transition from one state to the other. This would be the default state of how the brain works, even if we don't do anything to influence it,' Martijn says.

In theory, he agrees, better control over how the brain moves between these states could make all the difference to how life feels and to how efficiently the brain deals with a task. 'It's like a recipe: you have some ingredients ... a good chef could make a dish out of them, but a Michelin Star chef could cook up something just a little bit better. With that respect, I do think that having your brain be able to combine and link stuff together in a more efficient way would be beneficial for the higher brain functions that you mention.'

So I don't seem to be barking up totally the wrong tree with this idea, which is a relief because the other plan — to scan my brain, do an intervention, and then point and marvel at the bigger blob on a second scan — was never going to work. Even if it did, it wouldn't say much about how my brain was actually working differently afterwards. The only way to really watch my brain at work would be to take snapshots of brain activity over time and to watch the fluctuation of the various networks. Studies like this are, in theory, possible, but it is such early days — and activity probably varies so much from person to person that it would be inconclusive even if I had persuaded someone to do it to me.

Nevertheless, Martijn also agrees with my suggestion that getting into the right state by changing what you are doing should be possible. 'It could be that the brain state that you are currently in is not susceptible to reading, but if you go out for a run and you change the state of your brain then you might be slightly more susceptible to reading things. That's the boundaries that you can play with, and you can change your mindset.'

That's what I was doing in Boston when Joe and Mike suggested that the way to keep my attention on track was to 'embrace the fluctuations', or cycles of focus and mind-wandering in my brain. Until then, I had assumed that improving my ability to concentrate would involve strengthening the power of my frontal control networks, making them so strong that they could hang on tight to my powers of concentration until I was ready to let them go. Instead, learning to recognise both states and allowing them both to take a turn was the key to better attention control. The only way to learn this was to practise it until the state showed up, and then learn what it felt like. Having recognised this state, I now know that it also shows up when I'm swimming or doing yoga — anything that requires concentration, basically, but is also enjoyable and relaxing. Something similar was true for creativity: let the mind wander off the leash for a while, and you will probably come up with more creative ideas. You can bring the mind back to heel later to decide if what it brought back was any good.

Of course it would be nice if each of the states in table 4 mapped seamlessly onto a diagram of particular brain networks. Unfortunately, as with most things about the brain, it's not that simple. Each 'state' or zone is the result of a complex flow of information between many networks, some of them in the brain's well-connected club and some of them on the less connected outskirts. Mindfulness, for example, involves the default-mode network (during mind-wandering), the salience network (noticing that the mind has wandered), and the dorsal attention network (bringing attention back to the present). Each of these networks involves rich-club regions, but are not exclusively in the rich-club network. The cycling between states just keeps going until something more pressing comes up and takes your attention elsewhere.

## MAKE WAY FOR MINDLESSNESS

One thing that the hubs do seem to have in common is that they have a lot of contact with the default-mode network. This makes me think that embracing the default-mode network might be a powerful way to drive the brain in general. Recent research by Danielle Bassett and her team at the University of Pennsylvania has found that the central location of the default-mode network and its dense connections with the rest of the brain means that it is the least energy-hungry route from one brain state to another.[1] Which suggests that, rather than maligning mind-wandering as a problem state, perhaps it's time to rebrand it as just as important a state as focused attention. Mindfulness may be all the rage at the moment, but Bassett's work suggests that it is at least as important to make room for mindlessness. Without going through a spate of neutral, the brain might get stuck in one gear, and it'll take a lot more effort to get into another. It confirms my hunch that you can't be mindful all the time, and even if you could, it might not be the best state for everything that you want your brain to do.

Allowing what seem like undesirable states to have their say is a theme that is cropping up in other areas of psychology. I recently reported on the mental state of boredom,[2] which has only recently begun to attract a lot of attention from psychologists. Unlike distractibility, which is often seen as the curse of the hyper-connected age, boredom is, some researchers say, chronically underused nowadays. We have so many opportunities to distract ourselves from it — TV, smartphones, hilarious cat videos — that we don't have to suffer it for long. Sandi Mann, a boredom researcher at the University of Lancaster, reckons that this is a bad thing. If we never give ourselves the chance to be bored, we are robbing ourselves of not only much-needed down time but also the chance to let the mind wander aimlessly, and perhaps creatively.

Not every scientist who is studying boredom thinks that it should be encouraged — and, personally, I'm not entirely convinced that I need more of it, because, for me, boredom comes with a heavy dose of demotivation, which is anything but creative. But aimlessly mind-wandering, without the negative connotations, definitely has a place in any mental toolbox.

Anxiety, on the other hand, definitely does not. Stress can be a great motivator in small doses, but when it gets out of control, it is the enemy of the other mental states. All of them. Believe me, I have searched the scientific literature for an upside of chronic anxiety or a generally neurotic personality, and I just can't find any. There was one recent-ish study that found that the upside of worrying what people think of you is that you can't possibly be a psychopath,[3] but that's not actually one of the many things I was concerned about. And while some claim that anxiety is good if you're in a dangerous situation (because you're already primed to spot danger and can react quickly), by and large, it's a huge waste of energy. In my experiments, I found that anxiety was hijacking my focus, creativity, and my ability to think logically, and was playing with my perception of time. Basically, it was interfering with everything I was trying to do.

Thankfully, given the huge spanner anxiety puts into my mental and cognitive works, I feel this is a mental state that I have under much better control. Brain-wise, since emotional control comes under the umbrella of the prefrontal cortex, I'm guessing that I must have gained better control over the switch that gets me into a better-balanced state of mind. Perhaps the balance of signals from the frontal parts of the brain to the emotional centres has shifted so that the frontal regions have become better at getting their point across. Or perhaps fewer panicked messages are coming through, as a result of the cognitive-bias modification. It's probably a bit of both.

I'm reminded of a conversation I had with neuroscientist Oliver Robinson, from University College London, while we were both at the Society for Neuroscience meeting in Chicago. He gave a really interesting presentation about how the circuits that link the amygdala and the prefrontal cortex differ in useful anxiety versus pathological anxiety. The answer: in normal anxiety, the panic circuit is dialled up and down in a useful way. In problem anxiety, it gets stuck in the 'on' position.

Afterwards, I asked him which comes first in anxiety: does the amygdala react, and the prefrontal cortex has to then turn it down? Or does the strength of the PFC determine how strongly the amygdala reacts in the first place? Oliver's reply answered a lot of my questions about brain activity — it's a circuit; it doesn't really start anywhere. Activity is circling around between the two in a push-me-pull-you kind of relationship. If the circuit is always turned up to max, the activity in both parts of the circuit will be higher. No wonder it's so exhausting.

Prefrontal circuits are particularly high energy to run, it turns out. Another interesting outcome of Bassett's study is that while the prefrontal networks are absolutely crucial to mental control, their connections to other parts of the brain are actually fairly weak. The fact that they are out on a limb means that they need more energy input to engage (which is why it takes more effort to concentrate than to daydream). On the other hand, they are the gateway to harder-to-reach but important mental states, like calm control under pressure, impulse control, and the hard-working, pour-everything-into-solving-a-puzzle state I am calling effortful thinking, for want of a snappier term.

Bassett and her team point out that the importance of this weakly connected network goes against a major assumption in neuroscience — that the strongest connections are the most important. I would argue that the possibility that weakly connected areas are even more important than strongly connected ones also runs contrary to most laypeople's understanding of brain training. In this new view, strong connections linked by thick, insulated wires are not necessarily the most important for mind control. Sometimes, a weakly connected network can be incredibly powerful if you know how to use it.

I wouldn't say that I have gained total control over my frontal control networks, but I have certainly used them a lot this year. Another oft-ignored fact about mind control is that it always, always, involves control over attention — and that is a job of the prefrontal cortex.

Attention, and control of where you put it, is a major feature of the kind of brain that can put its mind to anything, and it kept cropping up in my various efforts to improve seemingly different things. An element of control is crucial in sustaining attention, to become aware of stress, to evaluate new ideas, to hold calculations in working memory in order to solve a problem, to manipulate a mental map of the environment, and to manipulate the passage of time as much as possible. It might be on the outskirts of the brain and not invited to the rich-club parties, but the frontal circuit is well worth the extra mental effort needed to keep everything else in check.

I'm aware that this sounds like a contradiction to what I have already said about there being one brain exercise that can boost the whole brain. And I stand by that — keep your money in your pocket and leave the 'brain training' apps alone for now. On the other hand, if, like me, you struggle to keep your mind on the job or to even pinpoint what in the world you are stressing about, then it seems that there is one exercise to strengthen prefrontal control — and it is meditation. Meditators, and people who are naturally mindful, show more activity in the prefrontal cortex and less in the amygdala, suggesting that they are better at controlling emotions. Expert meditators have a similar pattern. And it definitely helped me. I admit it: mindfulness helps you notice what is going on, and once you've noticed, more often than not you don't need to do anything about it. It's a bit like someone coming along and standing on the stress hose. After that, it just kind of splutters and stops. For me, once the penny dropped — that what felt like an amorphous cloud of unease was actually the result of one prevailing thought: *What if I'm doing this all wrong?* — it made the cause of my anxiety so much easier to tackle. I take back everything I said about meditation — if you only work on one state of mind, this one is the one to choose.

This probably isn't that surprising to anyone, given the high profile of mindfulness meditation these days. More surprising perhaps is the power of unconscious biases, and how they can only really be overriden using unconscious tactics. Finding that cognitive biases drag our conscious minds to see either the good or the bad side of life was a major 'aha' moment for me. So *that's* why I can't help feeling nervous in social situations, despite knowing that these are perfectly nice people who have no reason (yet) to think of me as an idiot. And *that's* why I can spot an accident waiting to happen even if it's hugely unlikely that it ever will. At some point in my life, my brain learnt that it was a good idea to keep an eye out for signs of danger, and it has been doing it on my behalf ever since. My genetic propensity towards a brain that is seemingly more plastic than average in emotional learning has made this lesson stick like glue.

This was another important lesson in my mission to take control of my mind. There are things that you can't just decide to change. In some people (like me, at least at the start of all this), an unconscious filter of attention is constantly seeking out things to fret about, which feeds the conscious mind a skewed version of reality. This means that the only thing the consciousness has to work with is the bad stuff, which is what makes negative biases (anxieties, worries, stresses) so difficult to think your way out of — the brain doesn't even get to see the alternative, happy point of view. It explains why you can't think yourself out of anxiety, and you can't think your way out of depression. It just doesn't work like that. If it did, no one would suffer from these things at all. It's not like cheering up or just not worrying hasn't crossed their mind.

In my experience, using cognitive-bias training as a prop did something much more profound than I initially thought possible. I know that, scientifically, the jury is out on whether it is ready to roll out on a large scale, but all I can say is that it worked for me. Slowly but surely, I feel like I have become less on the lookout for disapproval in people's faces, and less primed to see threats to life and limb all around. It was a highly specific lesson though — the anti-anxiety effect didn't generalise to work- or relationship-related anxiety. These neuroses had to wait until the mindfulness training kicked in. But the fact that the happy-face clicking of Elaine Fox's cognitive-bias modification training worked at all was amazing. All it took was a five-minute game that I could squeeze in whenever I needed a break from work, and it doubled-up as a way to practise the relaxed-and-ready zone. I quickly found that being in that state — not trying too hard, but paying just enough attention — made it much easier to find the happy faces. They just started jumping out at me and I didn't *have* to look at all. For me, it's been a win-win — so much so that I'm still playing the game every day.

I don't plan on stopping either, because if there is one thing we know about brain plasticity, it's that there is always the risk that things might slip back to baseline, especially if, like me, your brain has been set up to be more likely to react in a certain way. The idea that some people come with brains that specialise in some things and struggle with others sounds obvious, but it was something that I hadn't considered until I told Martijn how unhelpful I had found working-memory training. 'Well, maybe you aren't wired for that?' he suggested.

This brings up another truth about brain change that is often overlooked: you can't do anything you like. There might be some brain skills that I will never have, no matter how much effort I put in. And (this is the meditation talking) that is okay, too.

Martijn agrees. 'I like empowerment, but I think we should be realistic about the biological boundaries that we have,' he says. 'It might be that how the brain is wired and how the system is set up, you have the power to get better in some areas, but others remain locked away. It's good to be a dreamer, but you should be realistic.'

This is a conclusion that I have come around to in my navigation ability. I spent six weeks wearing a navigational belt and exploring the countryside, and learned, without doubt, where north is in and around my hometown. This experience has definitely helped me build a better mental map of my local area and also to learn to use things like the position of the sun and shape of the land to get my bearings. It revolutionised my ability to make cognitive maps in the areas that I used it in, but has it changed my brain's navigational equipment?

Personally, I have my doubts. By the time I got to Russell Epstein's lab, I had worn the belt religiously for weeks, and all the changes that could have happened should have at least started happening by then. Even so, I still performed way below average on all of the tests that involved making a cognitive map of my surroundings from scratch — and part of my brain's cognitive map-making system just didn't show up at all while I was there.

It wasn't only the scores — or indeed the brain-imaging results — that made me think that the necessary brain areas aren't properly online. It was the way that it felt, subjectively, when I was trying to solve the tasks they set me. Making and using a mental map didn't feel like a skill I that would probably improve with practice. It felt physically impossible, as if my brain couldn't do anything but flail around wildly, until it was time to guess.

It was interesting to see that activity in a part of the brain tasked with making cognitive maps didn't show up at all in the scanner, and that my hippocampus looked the exact opposite of the famous taxi drivers of London. And although I've been told time and again during this experiment that you can't tell anything meaningful about the brain by looking at one person (because there's too much potential for chance events to influence what you are seeing), it lines up well with both how I scored in the lab and struggle in real life.

This doesn't necessarily mean that it is impossible to override these problems, though. Brain plasticity alone might not do the trick, but what we definitely know about the brain is that it is nothing if not adaptable, and it is really good at learning and memory. The feelSpace belt didn't change my navigational brain, but helped store a map of my neighbourhood in my memory that I could draw on whenever needed. I used the belt to draw the map, and now it is there in my memory, hopefully for good. This is great news: you don't need to change the navigational circuitry in your brain if you can fill your memory with the information you need and then use it later. And even though the belt went back, I have a compass on my keys, so that if all else fails and I have a mental spin-out, I can always work out which way is home. If the sun is shining I might not even need to get it out of my pocket.

Knowing the limits of your brain can help you find ways around the blockages that are there, whether we like them or not. Human brains have the capacity to use tools for jobs that are beyond our natural capabilities — even, perhaps, to add a totally new sense such as magnetic-field detection. Why not use that ability to get around what is lacking? Martijn puts it better, I think. 'It might be that if you went back in the scanner that part of the brain still wouldn't light up, but you have been able to overcome that issue by using other resources to do the same function. If you don't have the number 5 you can still make it by adding 1 to 4.'

The power of this approach is that if you know what your brain is and isn't capable of, you can play to your strengths and adapt to your weaknesses using whatever props are necessary. That's why I think that one of the most important things that I have done this year is to get to know myself much better. Spending so much time filling in psychology questionnaires, doing behavioural and genetic tests, and getting brain scans has given me a much better understanding of the way my mind works, and the things it is good and not so good at. That's not easy, of course. As Benjamin Franklin wrote way back in 1750, 'There are three things extremely hard: steel, a diamond and to know one's self.' Science can help, though, which is why there are links to all the psychological tests you will ever need on the *Override* website. The first port of call for anyone wanting to change their brain is to find out what you are dealing with.

Be careful what you wish for, though, because there is no such thing as a free lunch when it comes to the brain. There are only so many resources to go around, so there is always the possibility that improving one skill might scupper your chances with another. I thought long and hard about whether to invest a lot of time and effort in logical-thought training, and in the end I decided that I didn't want to favour something that might take away from my naturally creative side. Too much control over my thought processes might take me towards being someone that I don't particularly want to be. And now that my emotional side is a little more in check, I feel like I'm quite logical enough, thank you.

Finally, there is another possibility that I can't ignore, because if

I do I will be hauled over hot coals by the neuro-police. It is possible that all of the changes I can feel are explained by something other than real change in the brain. They could also be partly or wholly down to either the placebo effect, regression to the mean, or the simple fact that when you do something more than once you get better at it. I have invested a lot of time, money, and energy into trying interventions that I hoped would improve my skills in key areas, and I chose things that I genuinely wanted to change about myself. Even with all the journalistic scepticism in the world, might my hopes and expectations be enough to make my wishes come true?

I can't deny that it's a possibility. In medical trials, even when patients are told they are being given an inert sugar pill, a large proportion of them report feeling better after taking it. It isn't well understood, yet, why this happens — it could be that there is something about being prescribed even a fake drug by someone in authority that stimulates the body to heal itself. Whatever it is, it must be something to do with the brain, because there is nothing in the pills that should fix anything.[4] Recent research has shown that the placebo effect affects the brain in real, physical ways, making it respond to pain and stress differently, as well as affecting memory. There is even interest in trying to harness the placebo effect for conditions that are difficult to treat any other way.

The same is probably true in psychology experiments, says Walter Boot, a psychologist at Florida State University, in Tallahassee. In a recent paper, he pointed out that experiments designed to test cognitive-training programmes almost never factor in what the volunteers expect to happen.[5] And this, he reckons, could be an important oversight. 'The potency of some reported placebo effects [in medicine] made us worry that similar effects might be operating in psychological interventions,' he told me in an email.

'We've found that expectation for improvement often matched actual improvement in these brain-training interventions. When this is true, it is difficult to attribute changes in performance to the intervention itself instead of placebo effects.'

It doesn't help that, in psychology, it is often difficult to disguise the placebo training as part of the real experiment, because it is fairly obvious that you are not in the training group if the game stays on the easiest level. And there is no getting away from the fact that in my experiments I had no chance of staying blind to what was supposed to be happening to my brain. I wasn't turning up at a lab to join in with a nameless scientific experiment that I knew nothing about — I was contacting specific researchers and asking them to do their best to change something of my choosing. The expectation effect is difficult to quantify, but as another of those powerful unconscious biases, it almost certainly muddies the waters in my experiments as well as in lots of proper, scientific ones.[6]

To me, though, the possibility that I have willed my brain to work better is an interesting possibility in itself. If choosing some areas to work on and dedicating some time to them is enough to make real, measurable changes in my brain and how I feel, that's fascinating too. Either way, my brain proved itself to be capable of change. I feel more in control, and with a few new skills that I didn't have a year ago. Statistical reliability can only come from properly controlled trials, but what I can say from a human point of view is that it was absolutely worth doing.

I started this project feeling slightly out of control, driven by glitches that seemed out of my power to change, and sceptical that I'd be able to do anything very much about it. I am very glad to have been proven wrong. I am finishing it feeling like a proper grown-up, with a mind that I know is absolutely capable of doing what it needs to, and with a better understanding of its strengths and shortcomings. It's been a fascinating — often gruelling, but always entertaining — journey into the workings of my own mind.

And now to the tricky part: turning this into something anyone else can use.

## A 20-MINUTE WORKOUT FOR THE BRAIN?

When I started all of this, I was looking for the brain equivalent of a jog around the block and 20 push-ups. The good news is there is one. The bad news is that it's a jog around the block and 20 push-ups.

I'm being facetious of course, but having read the studies, talked to the experts, and tried various brain interventions for myself, I have to conclude that anyone looking for one brain-game to rule them all is going to be disappointed. Most analyses of generic brain-training programmes so far suggest that any benefits are small, non-transferrable, and temporary.

Even the scientists who think that working-memory training can improve general intelligence in the lab point out that there is no evidence that it makes any difference to how the brain works in real life. So, for my money anyway, brain training as it is currently sold doesn't seem worth doing. You're better off exercising, which is known to put the brain into a physical and chemical state that primes it to learn, and eating a good balanced diet to make the brain as ready for the job as it's going to be. Even then you'll probably bump up against the genetic ceiling of your ability at some point.

It doesn't sound like it, but I reckon this is good news. Adding a one-size-fits-all brain-training routine to the list of things we should be doing every day (physical exercise, drinking eight glasses of water, eating five portions of fruit and veg) doesn't feel terribly helpful. It feels more like another stick to beat ourselves with.

On the plus side, picking specific skills and working on them is definitely possible. So if you want to make use of the plasticity of your own brain and feel the benefits almost straight away, my first bit of take-home advice is this: *pick what you want to work on*. There might not be a quick mental workout that will increase your general intelligence, but you can almost certainly get better at maths.

The second, when possible: practise these things in real life rather than on a disembodied app. Once I'd got rid of the panicky feeling and started to focus, the best intervention for improving my maths skills was to practise maths problems, starting with easy ones and working my way up to the harder ones. Similarly, there was no better way to practise not getting lost than to go out exploring — just with a couple of props as a safety net, in case my natural ability failed me. A similar benefit might have come from computer gaming, and the evidence looks convincing enough that it is definitely worth a try for anyone who is that way inclined or short on time to wander. But given the added brain benefits of being outdoors and getting exercise, practising the real thing is best, if at all possible.

Not everything can be practised directly, though — like those mental states I've been banging on about. This is where the advice starts to get more difficult to pass on without resorting to self-help-style platitudes. To be clear, I have never intended this to be used as a self-help book: I don't feel comfortable making sweeping statements about what everyone should be doing for their brain, when the experts have spent over a year telling me that no such thing exists.

Still, I am confident that there is enough evidence to say that it is worth trying to master the relaxed-and-ready zone.

Sadly, Joe DeGutis' training isn't available commercially. Having since tried meditation, though, I have realised that the feeling I get when doing an open monitoring meditation (alert to sounds, thoughts, and feelings, but not getting dragged off by them) is much the same, as is the feeling I get when swimming in a deserted pool or strolling through the woods. Despite all of my initial reservations, it does seem worth learning meditation, preferably from a real person who can advise you along the way. If you don't fancy that, it seems to me that if you can access that state then it doesn't really matter what you do: singing, climbing, running, playing an instrument — whatever floats your boat but isn't too easy or too all-encompassing.

There's no denying, though, that relaxed and ready is hard to describe in words in a way that gives someone else access to the same state. You have to feel it for yourself. If that sounds too vague for your liking (and I wouldn't blame you), one option is to try an EEG meditation aid of the type that is starting to appear on the market. They're not cheap, at around £200, but having test-driven one (the Muse headband) I found that it was a pretty good way of getting feedback about when you are and are not in that state. With these devices, the relaxed-and-ready state is measured via a relative dominance of alpha waves, which are well established as a marker of alert relaxation. While the Muse app doesn't tell you exactly what your brainwaves are doing, it does extract that information in the background and process it into a rating of calm, neutral, or active mental state. Muse's scientific director, Graeme Moffat, told me that the aim is to provide a set of 'training wheels' to learn meditation, so you know when you have got it.

However you get there, once you've got it, my advice is to practise that feeling whenever possible. Now I've got the hang of it, I use it for things that I don't feel like doing, such as: reading long-winded scientific papers; doing chores; playing Lego Batman when I'd rather be reading a book; enduring long presentations at conferences. Anywhere where my natural state would be distracted and stressed — having an alternative state of mind to slip into has been a revelation, and is well worth all the sitting around.

Another option is to cheat by loading your attention system to just the right extent, as in Nilli Lavie's load theory. There are a few apps and websites that do this for you, in case you don't have the time to colour your pages and find just the right level of atmospheric noise. I have to say that, when I tried them I found the musical choices (ranging from plinky-plonky to new-age) quite irritating, but enough people have recommended them to me that I guess they work for some. Ommwriter.com is about the best I found, and it has the added feature of being able to change the background colour as well as the ambient noise, and it gives off a strangely satisfying clicky noise when you type. I wrote part of this chapter with it, and it's really rather nice.

Sometimes, though, relaxed and ready isn't quite a focused enough state of mind for the task at hand, and only a short, sharp session of effortful thinking will do the job. This is a tricky one to advise on because of the conflicting evidence about whether or not training cognitive control (all those prefrontal cortex skills) transfers past certain tasks to make a generally better-controlled brain. The answer to that is: probably not. Nevertheless, I do feel that I have a better ability to make myself think hard when necessary now; I just can't put my finger on exactly what was the defining intervention, or if indeed, there was one. If I had to put money on it, I'd say it had a lot to do with reducing anxiety and learning the relaxed-and-ready zone. Both freed up mental resources that left more in the tank when I have to think hard. Can I offer a one-size-fits-all bit of advice on how to do it? I'm not so sure.

One possible shortcut is to use natural variations in alertness throughout the day, which seems to link into how our natural circadian rhythms affect body temperature. Any time the core body temperature dips below 37 degrees Celsius, concentration suffers, which means that, by this measure, first thing in the morning is the worst time to try and concentrate, and peak periods, when it'll be easier, come between 10 am and noon, and 3 pm and 6 pm. There is an easy override for this, though: increase body temperature with some exercise or a hot shower and focus should follow shortly, at least for a while (see graph below).

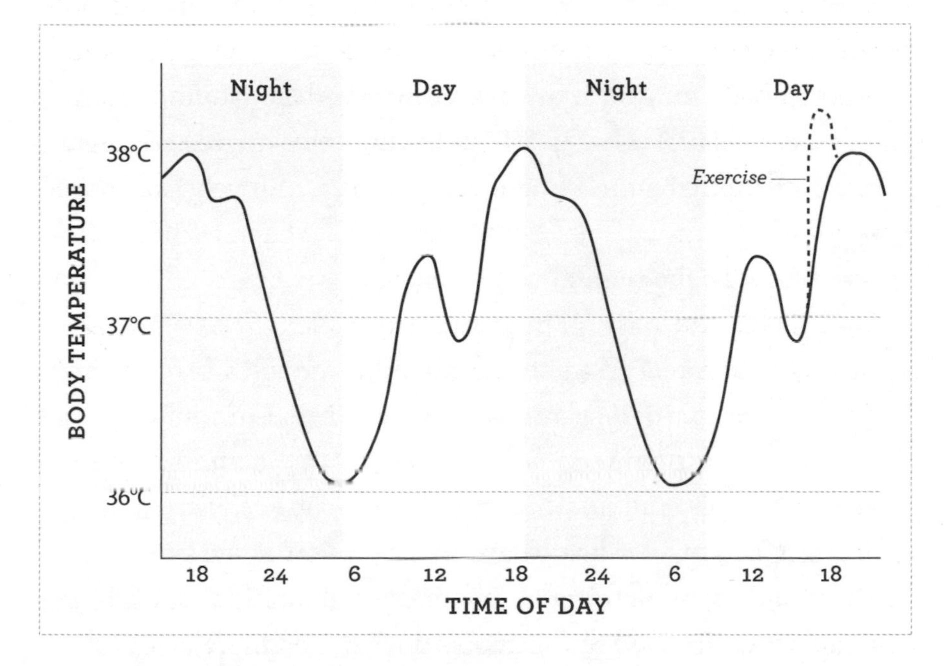


Fig 19. Body temperature variation over two days. (Adapted from: Gerry Wyder)

Then there is the all-important antidote to concentration: the much underrated state of mindlessness.

Zoning out might not sound like mental exercise at all, but if Danielle Bassett's work on the importance of the default-mode network is anything to go by, then remembering to make room for aimless mind-wandering is about the most important thing you can do to get the best out of your brain.

Going mindless doesn't necessarily have to mean a long period of staring out the window. If your day is back-to-back hectic meetings, maybe take the long route to the next room while you let the mind skip off wherever it likes. It might involve imagining what you'd really like to say at the next meeting, or pondering where you'd fly to tomorrow if you had no ties and limitless cash. It doesn't matter, as long as you disengage the busy thinking parts of the brain to give it a rest before diving back into professional mode. Let it go for as long as you can before the adult world drags you back in. And if anyone comments that you are staring into space, you can always tell them that you are practising an important mental state known as hypofrontality — that should shut them up ...

And, finally, the mental state closest to my heart, and the one I'm most glad to have gained override privileges for: anxiety. Clearly, it's to be kept on a short leash, however possible. Cognitive-bias modification helped me with social-threat processing, and seems to have made me less on the lookout for doom in general. Past that, it was *meditation, baby*, all the way. I take back all my initial scepticism, because I have to admit that it works for me. I wish I could say something more concrete about this, but I'd have to stray too far from what is actually known, and there is too much flannel out there already. All I can say is that the interventions I tried in chapter two worked for me, and I hope they are of use to someone else out there, too. Chronic anxiety is not necessarily part of you, just a glitch in the system, and one that doesn't always have to run the show.

In summary, I've said it before but it seems worth repeating: much of what we need to run our mental lives is on-board already.

The trick is to be able to choose to use it. And, yes, you can override the basic features of your brain.

Try these:

1. Exercise — preferably outdoors
2. Seek out a mentally challenging job or hobby
3. Do a course in mindfulness meditation
4. Remember to let the mind wander to reset
5. Pick your skills and practise them in real life
6. Hang in there, because a better understanding of the brain, and more direct ways to override it, are on their way ...

CHAPTER 8

# The Road from Here

*'I don't know where I'm going from here, but I promise I won't bore you.'*

— DAVID BOWIE

It's no exaggeration to say that the changes in my mental control ability and cognitive skillset over the past year or so have been life-changing. But they are also miniscule compared with what will be possible in the future.

We are teetering on the brink of a new world where it will be possible not only to plug in to your own brain and read what it is up to but also to use that activity to move *directly* into the most useful state of mind for the job. None of this mucking about with trial and error — all the information you need will be right there to use as you see fit. And if the brain is stubbornly refusing to play ball, or is too stuck in its ways to be nudged out of old habits, it might well be possible to give it a gentle zap to get things moving in the right direction. If that isn't enough for the brain control-freak of the future, there is also the exciting possibility of adding to the brain's information base by bolting on new senses and even huge data streams for the brain to integrate without any conscious effort.

All of this might sound far-fetched, but it really isn't. I have experienced versions of all of these things in the lab this year and almost all of them at home, too (I could have done them all, but Roi didn't make home-zapping sound like a great idea). True, not all of these technologies are ready for prime time yet, but it won't be too long before they are. Chances are my grandchildren will look back on this experiment of mine as really rather quaint.

We are already making some progress towards more direct forms of brain override, through the world of home EEG. For around £200, you can buy a home-EEG headset that is, in theory, capable of tracking your brain's electrical signals in real time and turning it into a read-out of your brain as it moves from focused to daydreaming to meditative to asleep. At least on the one I tried — the Muse headband — this tell-all read-out doesn't come as standard at the moment; the Muse app translates the raw data into three categories: calm, neutral, and active. Muse's scientific director, Graeme Moffat, wouldn't tell me what each category represents because the algorithms they use are a trade secret, but he did say that, if I wanted to know more, I could download a second app called Muse Monitor that does provide the raw data. I used Muse Monitor to do a 10-minute guided meditation while recording my brainwaves and ... with such a cacophony of data, my computer crashed — so that was inconclusive. In theory, though, with a little patience and a computer that can handle it, it is possible to look at your brain's general pattern of activity over time and to relate that to how different patterns of activity feel in real life.

To be fair, tracking raw brain activity in real time isn't really what these home headbands are designed to do: the main point of them is to measure and then tweak certain brain signals, with the aim of either changing your state of mind or controlling an external device linked to a computer. Some people have hacked these devices to drive electric wheelchairs, for example — and there are even a few toys on the market, including one that uses EEG and the power of concentration to fly helicopters.[1]

These devices all work on a much simpler version of the kind of neurofeedback that has been in labs for years and that has shown some promise in treating ADHD[2] and PTSD.[3]

As you might expect, what is available commercially now is pretty simplistic. Most of them only offer two states: 'concentrate hard' or 'relax and let go'. In addition, home-EEG kits, at the moment, only record from a tiny number of electrodes — four in the case of the Muse, one in some of its competitors. In the lab, researchers more often use caps with 64, 128, or 256 electrodes. The more electrodes there are, the better the chances of working out where the signal is coming from, and using that information to change your brainwaves accordingly. With home versions now, we can only get a broad view of which state is dominant overall — which is great, but if future versions of the technology involved more electrodes, positioned all over the brain, it might be possible to target particular brain regions and the circuits they are a part of. It's easy to imagine future versions helping children in class to practise controlling their attention for longer, using an EEG-controlled toy; or for classroom assistants to be able to monitor children with ADHD and intervene when their attention has drifted (and before the child has resorted to bad behaviour to get the stimulation they need).

In the future, with more electrodes and more research, there may be options for a far broader range of training than is available now — perhaps to enhance cognitive control over things like emotions, working memory, and attention.

One avenue of research, still in the lab at the moment, aims to boost this more general kind of cognitive control by targeting one particular brainwave band — theta — specifically in the top and centre of the prefrontal cortex. High levels of theta brainwaves in these regions has been linked to better executive control and to lower levels of anxiety. Recent studies seem to show that it's possible to increase theta brainwaves to the right level for cognitive control, and for this to improve performance on the kinds of working-memory and attention tests I have been doing so many of recently. With cognitive training for working memory so much up in the air, the prospect of being able to cut out the middleman is one area to watch in the future.[4]

If neurofeedback could be designed for use in training mental flexibility, that would be even better. Intriguing studies by Bernhard Hommel at Leiden University, in the Netherlands, have experimented with riding the balance between frontal control and flexibility elsewhere by first boosting gamma brainwaves in the frontal region with neurofeedback and then using meditation to encourage attentional flexibility. If this ever becomes a reality outside of the lab, it sounds like a potential route to the kind of controlled flexibility that I have been trying to master.

I had a try on a lab set-up, with four times as many electrodes as the Muse, while visiting Klaus Gramann's lab in Berlin. They offered it to me as a kind of booby prize, to make up for the experiment I had come for not being ready. Perhaps they felt they had to plug me in to *some* kind of wiring to make the trip seem worthwhile. At the time, I was a bit disappointed, but it actually turned out to be pretty fascinating.

The downside of lab neurofeedback is that it usually involves squirting conductive gel onto the skin where the electrodes will sit, to help the weak electrical signals from the brain to get through. One of Klaus' students, Cao, attached the cap and electrodes to my head, while another, Laurens, filled a syringe with bright green, gloopy gel. 'Don't worry, we have facilities for you to wash your hair afterwards,' Laurens told me. 'Well, we have a tap ...'

The first step in neurofeedback is to teach the computer about your own personal brainwaves. Each person's brain is slightly different — and the way it is folded varies quite considerably — so there is no one-size-fits-all brainwave pattern. Then, once the computer has gotten a measure of me, I will turn the tables and use my mind to control it with my thoughts. This particular training involves switching between deliberate thinking (rapid-fire sums that popped onto the screen) and happy and relaxing thoughts (whenever the screen changed to floating stars). When the stars appeared on the screen, I visualised snuggling up under the duvet with my family on a Sunday morning. This back and forth between states goes on for ten minutes or so, and it is surprisingly hard work to switch back and forth, on command, between thinking hard and chilling out. After a quick look at the data to make sure they picked up a good enough signal, they let me try to affect what was happening on the screen by changing my mental state. If I thought hard, the screen went blank. If I went to my happy place, stars began circling on the screen. As soon as I looked at the stars and thought about what they meant, they disappeared and the screen once again went blank.

With a bit of practice, it's actually quite easy to switch between the two states — so much so that it's easy to imagine this being built into some kind of creativity-enhancing screensaver.

The Muse headset, and others like it, is designed to give you similar control over your brainwaves, except it's more geared towards sustaining calm rather than focus. Hardcore meditators might be less than impressed by the idea that you need a machine for this, when you can actually just sit still and focus on your breath for free. One thing that does bug me about the Muse app, I admit, is the constant feedback about whether you are doing it right. Gill, my meditation teacher, spent a lot of time explaining how the point of mindfulness is to be calmly aware of the moment *without judgement* — 'Without worrying if you are doing it right,' as her practice CD says over and over again. The Muse app system of feedback seems to go against this — it's hard to *not* feel like you're doing it wrong when a storm starts to whip along a previously calm and peaceful beach. Then there is the fact that it takes a couple of minutes to calibrate the app to your current brainwave pattern before you can even start. A couple of times, I couldn't even get it to synch to my phone and gave up, feeling more stressed than when I started. But still, for people who don't like the new-age connotations of meditation, thinking of it more as tech-assisted mind-control might be easier to stomach and just as useful.

The good news is that regulating certain brainwaves in neurofeedback does actually change the brain in real and useful ways. A recent study found that, after just 30 minutes of alpha-wave-based control, there was an increase in functional connectivity in the salience network (the one that spots your mind going off track). Something about learning to engage that network made it function better, which bodes well for applying this kind of learned control to other parts of the brain.[5]

A newer version of neurofeedback research is even further reaching. Until fairly recently, EEG has been the only way to track changes in the brain in real time. Using EEG alone, though, makes it difficult to pinpoint exactly where in the brain the signals are coming from. Functional MRI (fMRI) has the benefit of being able to show exactly which parts of the brain are active at any one time, but it is much slower — making it difficult to use for neurofeedback. Recent developments in real-time fMRI, though, have made it possible to track brain activity as it happens. Studies using this technology have found that it can be used for neurofeedback: if you show volunteers their own brain activity and give them some time to practise, they can learn to consciously drive up activity in particular brain regions.

The cool thing about this technology is that it has shown that, when a person learns to control activity in one brain area, such as the insula or amygdala, it not only boosts the brain region they are working on but, as you might expect, it improves connectivity throughout the rest of the network, too. Something similar has been found for the prefrontal cortex's links with the inferior frontal gyrus (an area associated with the 'flow' state) after just a few training sessions.[6]

The other big advantage of fMRI over EEG is that it can track changes in any part of the brain — whereas EEG struggles to pick up a signal from areas deeper than the wrinkly outer cortex; lots of interesting things happen deep in the brain, and it is interesting to find that these can be targeted directly, too. Of course there is a considerable downside: you need a huge MRI machine rather than the few electrodes and wires required for an EEG, so it's nowhere near being useable at home, but for research purposes — to see what is physically possible through neurofeedback — it's proving very interesting.

The hope is that this kind of technology could one day be used to help people with brain or psychological disorders to 'normalise' the workings of their brain; and eventually to give anyone who wants to improve their brain function a good shot at selecting and working on particular skills. It's a way off yet, but it's there, being fine-tuned every day, informing the development of technologies that, one day, anyone could use. A recent review of the technique concluded that there are plenty of unanswered questions, including whether activity can be dialled down as well as up and whether people can continue to change their brain's activity outside of the scanner.[7] If the answer to the second question turns out to be 'no', then it's a dead end for most of us unless someone makes MRI scanners considerably smaller and cheaper. On the other hand, as these new states are recognisable to the person experiencing them, and teachable by other means too, it opens the possibility of training someone to work their brain in the right direction and then sending them home to carry on with it indefinitely.

Another development that sounds way more sinister is the even newer world of covert neurofeedback. Weirdly, it seems that you don't even need to be part of the equation while your brain's activity is being altered. At the 2015 Society for Neuroscience conference in Chicago, Michal Ramot, of the National Institute of Mental Health, in Bethesda, presented data from her recent experiments in which volunteers learned to change neural activity in two areas of their visual cortex (at the back of the brain). When playing a game in the scanner, volunteers knew that they would earn real money for giving the right answer, but they didn't know what the right answer actually was. Moreover, they were told that the feedback they were getting was random. Nevertheless, the volunteers were able to learn how to earn money in the game without being able to say what they had learned or how. At the conference, Ramot suggested that a similar approach could be used to treat mental disorders, to help people learn new skills without having to try too hard, or to help locked-in people communicate with the outside world. Call me pessimistic, but my thoughts immediately went towards imposed mind control and the myriad ways that marketeers and, worse, governments, might be able to use the technology against our will.

Whichever way it's done, neurofeedback is going to involve a certain investment of time and effort. If that isn't your bag, then perhaps it's worth waiting for home brain-stimulation to hit the market — preferably in a safe, properly regulated way. Using electricity to 'zap yourself better' has actually been around since the first century AD, when the Roman physician Scribonius Largus persuaded headache and gout sufferers to let him put electric fish on whichever end was hurting (apparently it made the affected area go numb for a while). And for all the health warnings, when done properly, tDCS does seem to drive real changes in the brain, and real changes in behaviour, at least in the short term. I have seen the benefits twice now in the lab and, caveats aside, both times the size of the effect was pretty impressive.

Roi Cohen Kadosh feels the same, and he has seen a lot more data than I have. There are even more promising results from studies where a different kind of stimulation is employed — transcranial alternating current stimulation (tACS) — which can be used to select particular brain-frequency bands that best suit the task at hand. 'There are some very impressive findings, in my view,' Roi says, citing a recent paper from his lab in which they found that tACS boosted gamma-frequency brainwaves in the prefrontal cortex; the speed at which people were able to answer logic questions increased by several seconds when they were stimulated in this way.[8] This is the kind of improvement that could make a big difference in a test situation, he says: 'In an IQ test, you are talking about being able to get through more items, and therefore the score is higher.' The only catch is that no one knows yet whether stimulation would change ability in the long term or just for an hour or so after stimulation. It could be that you'd have to use these tools selectively, before an important test, interview, or assessment, which brings up all kinds of ethical dilemmas about fairness. Or perhaps everyone will be doing it in the not-too-distant future.

This reminds me of a conversation I had with Lila Chrysikou in Kansas, as we headed back to the lab after my morning in the scanner. We were both flagging, and shared a caffeinated chocolate bar that boasted on the wrapper that it contained as much caffeine as a cup of coffee. Maybe in the future we'll be able to zap our brains to wake up, instead, I suggested. 'Maybe,' she replied. 'It's the same principle ... just more direct.'

When it comes to directly stimulating the brain, there are a lot of things still to iron out in the lab — not least finding out what exactly it is doing to the brain, but also which types of stimulation work best for particular cognitive abilities, what kind of people benefit the most, and what the trade-offs might be in boosting one part of a very complex machine while potentially doing harm to another. And, more importantly, someone has to bite the bullet and test whether the technology is safe to use as often as we might like.

One way that this could be achieved is if home tDCS-ers volunteered to participate in a study of long-term safety. If the experiments were done on a group of people who were using the technology anyway, the ethical concerns would presumably be less. The problem is that, at the moment, the two communities are not talking to each other. Amar Sarkar, who zapped away my maths-related fears in Oxford, believes it's time for that conversation to start: 'The people doing brain stimulation at home are driving the market,' he says. 'They are being condemned by scientists but the scientists are not communicating with them.' On the other hand, the neuro-hacking community is less than impressed with the scientists' warnings over safety. 'They say that we sit in our ivory towers and publish for each other. But both sides have the goal of making this available if it works.'

It will take a while to mend these bridges, and to get real answers to what the scientists — and the rest of us — want to know. The likelihood is that future studies will in turn throw up more questions as it becomes clear just how different everybody's brain is in how it reacts to stimulation. There are similar issues with all brain-changing technologies, though — and indeed in standard medicine. It is becoming increasingly obvious that not all drugs work well for all people, and that ideally we need personalised medicine, tailored to each person's individual needs. In the same way, just because I found tDCS to be useful and working-memory training to be no help at all, it doesn't mean that the same would be true for everyone. With tDCS, the subjective feeling of being stimulated is different for everyone — in Kansas, I felt zoned out and wonky, but Lila told me that some people don't feel anything at all. She also told me that one volunteer got so freaked out by the whole idea of brain stimulation that he fainted before the current was even switched on.

Susanne Jaeggi, the main researcher in the pro-working-memory training camp, says that this is her main beef with the idea that cognitive training can be thrown out before it has even got going. 'It's not one-size-fits-all — if you have major depression, for some people, cognitive behavioural therapy is the way to go, for others, psychodynamics works better ... I think the same is true for cognitive functioning. For one person, targeting working memory is the thing to do that they like, but for another person it might be tDCS, or for another person it might be musical training or learning a new language or practising impulse control skills or doing mindfulness meditation. I don't think we should think about it as "everyone should do the same thing" — that's not the way we work,' she told me. 'You wouldn't tell someone wanting to get fit that the only way to do it was running,' she adds — they might prefer swimming or cycling or dancing, and that would work better for them. 'It has to be something that you like, that is tailored to your personality as well. That is where I think we are not there yet. That's what we are currently working on. Ask me again in ten years, probably.'

With all of these caveats in mind, the one thing that I did find in all of the neuroscience and psychology labs that I visited was a huge amount of excitement about what should be possible in the future.

There are other tools in the mix, which I haven't personally experienced. One of these — vagus-nerve stimulation — sounds particularly interesting for the future. The vagus nerve meanders through the body, linking the brain to every organ in the body, and has a huge number of branches in the gut. It's part of the information highway by which the body and the brain communicate with each other. The main message that the vagus nerve sends is: calm down. After a bout of stress — which has the heart rate, breathing, and inflammation going — activity in the vagus nerve reverses all of those changes and puts things back to a resting state.

Vagus-nerve stimulation has been used to treat epilepsy since the 1990s, via electrodes implanted in the neck — the idea being that stimulating the 'calm down' response in the brain will help when a seizure is building. It is also being used as a last-resort therapy for people with depression who have had no relief from any other treatment; and, because of its ability to suppress the inflammatory part of the immune system, it has been used to provide relief from rheumatoid arthritis, an autoimmune disorder that causes painful inflammation in the joints. Other potential applications include other inflammatory diseases, like Crohn's disease, as well as migraine and chronic pain.

So we know that electrically stimulating the vagus nerve calms the body down. We also know that people in general vary in the strength of their vagus-nerve response — known as vagal tone. This, by the way, is why some people are able to think rationally in the midst of a crisis while others are running around, panicking and looking for the door. These two things together have led some to speculate that gaining better control over our vagus nerve would be a good idea for all of us.

Having electrodes inserted into your neck is clearly quite extreme, but recently a company has gained approval in the UK, Canada, Australia, Germany, and Italy to trial a non-invasive vagus-nerve stimulator for migraines and cluster headaches. The stimulator fits in the palm of your hand and can be used for a couple of minutes, two or three times a day. The device is still being tested, and isn't available outside of clinics yet — but it's not out of the realms of possibility that one day we could slip a handheld vagus-nerve stimulator into a bag and use it before a big interview, or when work is getting a little too much, or when a migraine threatens to pop up and ruin the day. It's one to watch, that's for sure — and there is yet more evidence that if you want to control the brain, you should take a good look at what is happening in the rest of the body, too.

Also on the subject of altering the body to affect the mind, the most unexpected brain control tool that I came across in my experiments made it possible to bolt a totally new sense onto the body, and for my brain to translate that information into something it can use. The weirdest — and coolest — thing about the feelSpace belt was that it was possible to take a completely unnatural sensation (buzzing on the waist) and attach new meaning to it: in this case, a sense of magnetic north. With this information, it was possible to do things I couldn't do with my brain alone — such as develop a far more accurate mental map of my hometown, helpfully aligned to magnetic north.

My experience with the feelSpace belt got me wondering what else might be possible to add onto the human brain. On one level, the idea of supplementing our natural senses is not that different to using night vision goggles to turn infrared light — which is outside of the natural limits of human vision — into something we can see. Or a bat detector, which picks up bats' ultrasound calls and brings them down a few notches into an audible set of clicks. The difference between slightly altering the frequency of sounds we can hear and what something like the feelSpace does is that the feelSpace adds a sense that no human has any built-in version of — feeling the magnetic field — and it does this by hijacking another sense designed for something else entirely.

Actually, researchers have been adding senses by co-opting spare bits of skin for several decades, at least in the lab. As long ago as 1969, researchers turned visual information into a physical sensation on the upper back, which blind people were able to use as a substitute for vision. Neuroscientist David Eagleman, of Stanford University, is currently doing something similar for deaf people: transmitting sound, via an intricate pattern of vibrations, into a smart vest. It works so well that deaf people have been able to use the technology to decode human speech in real time.

Using the skin to bolt on new skills makes perfect sense. When you think about it, we have a lot of the stuff, it is packed with sensory neurons and, most of the time, is sitting under our clothes not doing very much. If we could use those sensory neurons for something else, it could take us way beyond the senses that we naturally have on board. Emergency rescue workers, for example, could be fitted with a vest that was linked to an infrared camera, enabling them to feel where earthquake survivors might be buried under the rubble, using their body-heat signature. Or, perhaps, a chemical sensor, plugged into some kind of tactile band, could detect subtle subconscious signals of fear, arousal, or comfort in the sweat of people around you.

Eagleman is thinking beyond just adding senses: he is currently trying to use the vest to represent huge amounts of data — for example, from the stock market — that people could learn to understand and react to, quickly and subconsciously. He has even floated the idea of a couples' version, where you could keep track of your spouse's emotional state in real time. (Personally, I'm not sure that is such a great idea.)

I arrange a chat with Eagleman to find out what he thinks the limits to all this are. I manage to secure a slot of five minutes as Eagleman drives to the airport to give one of many talks about his research. Thanks to his popular books, TV shows, and his 2015 TED talk, Eagleman is a busy man — he is, as *The Times* newspaper put it in the days before we talk, the 'rock star of neuroscience'. I think what they mean by that is that he is enthusiastic about the future of neuroscience, can explain it in a way that other people can understand, and seems like an all-round nice bloke. Actually, I've met a lot of people like that this year — when you take the time to go and talk to neuroscientists, it turns out that they are normal people like the rest of us and have lots of interesting things to say. Not all of them have the fireman's hose of enthusiasm that Eagleman does, but still, it's a field packed with very interesting people indeed, and I've had a great time hanging out with them.

After checking that he's on speakerphone while he is driving (I don't want to be responsible for killing off the rock star of neuroscience), I ask him to explain how the brain goes about integrating a new sense. With the feelSpace belt, I knew that the buzzing indicated magnetic north, and I could then overlay that onto other knowledge of my surroundings, like the position of the river. Now I know not only where the river is but also its position in relation to north. How on earth do you go about making sense of hundreds of vibrations which could, at the beginning, stand for anything?

'Everything in the brain is about cross-correlating the senses and being able to compare one data input that it gets with another data input that it is trying to figure out,' he tells me. It's not magic: the wearer has to know what kind of information is encoded in the buzzes or it won't mean anything. The deaf volunteers aren't told the content of the words, for example, but they know that the buzzing represents language. Speech is actually a fairly easy one to learn this way, he says, because deaf people already feel their own speech as a vibration in the body. In principle, though, the brain's ability to learn a totally new input is limitless, so long as it has something to compare the new input to.

He also gives me the example of a drone pilot who learned to fly a drone better by wearing a vest that translates heading, pitch, and yaw into sensory vibrations on the body. The pilot can only put the sensations on his body together with what is happening to the plane by practising with both. 'When it's close up, he can see that *when the heading changes I feel this* and *when it rolls I feel that*,' says Eagleman. Then, 'when the thing is farther or out of his sight or it's night time, he now knows what the correlation is between these different data streams.' This, he adds, is how a baby learns to work out that banging something it can see and hold makes a sound. The baby bashes objects onto surfaces, drops them, throws them, and eventually works out how sight, sound, and touch all fit together.

Like the baby, who is working out what fits with what, it might be possible to add on pretty much whatever we like this way. One of Eagleman's favourite things to say is that the brain is locked in a dark box, and all it has to work with are electrical signals from the rest of the body. If it gets new information via the body, it doesn't care where that comes from, it just figures out what to do with it. This suggests that there is no limit to what our brains could use, if only we can bolt on the right sensors.

'One of my main interests is expanding our perception so we're seeing infrared and ultraviolet and we're hearing ultrasonic noise. The world is full of all kinds of interesting signals.' He points out that we are aware of a miniscule part of the spectrum of light that is around us. 'Even if we add on infrared and ultraviolet, we are still essentially at one ten-trillionths of it. So one of my interests is all these other giant parts of the spectrum that we don't see at all, but they are light, just not visible light.'

When we talk, the vest is in the final stages of design and planned for commercial release in late 2016. After that, he says, it's over to the hive mind to come up with new and interesting uses for it. 'That's going to be a big research project undertaken by thousands of people to just walk around for a while tapped into some piece of the electromagnetic spectrum and just figure out in their daily life what that gives them.'

The vest comes with an open programmer's interface so that anyone who wants to try it out on a different data stream, and has the skills to do so, can. 'We've thought of lots of interesting things, pilots, astronauts, prosthetic legs ... But we haven't even thought about what we haven't thought of yet. So there are lots of interesting opportunities for people to try data streams,' says Eagleman.

As for whether there are any limits to what can be added to the brain, I guess we'll see. There is only so much real estate inside the locked box after all, I point out. Eagleman doesn't think we're going to run out of room anytime soon, though. In a similar way to other areas of neuroscience that are starting to think of the brain less as a patchwork of areas with a particular job and more as an intricate wiring diagram, if you think of the way the brain stores information as a network, then the possibilities are endless.

'Medical students often worry that "if I learn one more thing, something is going to fall out",' says Eagleman. 'But, in fact, it doesn't work that way, because you start putting things into bigger patterns — like, "Oh, I see, this is the same as that, and these are all examples of a bigger principle." These are all ways that your brain saves a heck of a lot of room and energy, because it gathers the bigger picture.'

Which means that you might not need brain space to store new sensory information, you can store it as learned factual knowledge instead. The upshot of all of this, and in fact everything that I have explored so far, is that the brain might be capable of things that we haven't even begun to consider let alone try out in real life.

As an adult in 2016, I have been able to tweak the workings of the brain that was built through my genetic inheritance and life so far, and get it into much better shape for the life I now live. If these technologies, and the ones that no one has thought of yet, reach the masses, though, mine may be the last generation that will have to wonder whether we can improve on what nature and experience has provided. Whether you have to put up with your adult brain, warts and all, will cease to be a relevant question when you can zap, train, or otherwise whisk away the bad stuff while bolting on as many optional extras as you like. For now, though, what I can tell you for sure is that the human brain is an amazing thing to play with — and, no matter what your plans for it might be, it can almost certainly do more than you think.

# Acknowledgements

So many people have helped this project along the way that I barely know where to start. I do know that without Richard Fisher, editor at BBC Future, who commissioned the article on attention that led to the proposal for this book, the rest of it might never have happened. So thank you, Richard, for a crucial vote of confidence that got the ball rolling.

I am, of course, hugely grateful to the many scientists and their students who gave up their time, labs, and resources to help me experiment on myself, and who endured long periods of questioning and requests for information and research papers. Particular thanks go to Joe DeGutis and Mike Esterman at the Boston Attention and Learning Lab, Elaine Fox and her research team at Oxford University, Ernst Koster at Ghent University, Lila Chrysikou at the University of Kansas, Susan Wache, Peter König and the feelSpace team, Klaus Gramann at TU Berlin, Russell Epstein, Steve Marchette and the team at UPenn, John Wearden at Keele University, Marc Wittmann at the Institute for Frontier Areas in Psychology and Mental Health in Freiburg, and Amar Sarkar and Roi Cohen Kadosh at Oxford University for their help with my experiments. I had an absolute ball with you and I hope that I have done justice to your research. Thanks, too, to Heidi Johansen-Berg at Oxford University, Martijn van den Heuvel from Utrecht University, Sara Lazar at Harvard

University, and David Eagleman of Stanford University for some illuminating conversations along the way.

I feel very fortunate to have so many lovely friends around the world who were willing to put me up on my many research trips and allowed me to bang on about brains of an evening. To the Gosline-Knights in Boston, Neil Calderwood and Jessilyn Yoo in Berlin, Jolyon Bennett and Joanna Haigh in Oxford, and Valerie Jamieson in Didcot, thank you so much for having me, feeding me, and plying me with wine on the evenings when I wasn't having my brain zapped the next day.

Thanks also to my lovely, patient, and supportive friends and family at home for putting up with me, from the planning stages to the final written word. Special mentions must go to Cate for an incredibly well-timed mug purchase, and to Sally, Sole, Vanessa, and Ness for dangling cocktails in front of me at strategic points in the writing process.

I'd also like to thank Jeannie Campbell for sharing her experiences with me and for many fascinating discussions about her experience of rebuilding her brain. I am also grateful to Gill Johnson for introducing me to meditation 'without worrying if you are doing it right'.

I am hugely indebted to Peter Tallack and Tisse Takagi at the Science Factory, for believing in the project from the start and for brokering the deal that got the book written. To all at Scribe, particularly Philip Gwyn Jones and Lesley Halm, thank you for the opportunity to do this, and for asking all the right questions at all the right times.

And finally, to my wonderful husband, Jon. Without your support, encouragement, and incredible generosity with your air miles, none of this would have been possible. Thank you for your faith in me and your patience in allowing me what amounted to a second maternity leave. And to Sam, for proudly telling everyone that mummy is an author, even when I didn't feel like one: Thank you. I love you both.

# Notes

## Introduction


1 *The Discourses of Epictetus*, book 3, chapter 23.

2 'Lumosity to Pay $2 Million to Settle FTC Deceptive Advertising Charges for Its "Brain Training" Program', Federal Trade Commission, 5 January 2016: https://www.ftc.gov/news-events/press-releases/2016/01/lumosity-pay-2-million-settle-ftc-deceptive-advertising-charges

3 Owen AM et al., (2010) 'Putting Brain Training to the Test', *Nature*, vol. 465, pp. 775–78.

4 'A Consensus on the Brain Training Industry from the Scientific Community', Stanford Center on Longevity, 20 October 2014: http://longevity3.stanford.edu/blog/2014/10/15/the-consensus-on-the-brain-training-industry-from-the-scientific-community-2/

5 Shapiro DH, (1992) 'Adverse Effects of Meditation: a preliminary investigation of long-term meditators', *International Journal of Psychosomatics*, vol. 39, pp. 62–67.

6 Creswell JD et al., (2014) 'Brief Mindfulness Meditation Training Alters Psychological and Neuroendocrine Responses to Social Evaluative Stress', *Psychoneuroendocrinology*, vol. 44, pp. 1–12.

7 Arem H et al., (2015) 'Leisure Time Physical Activity and Mortality: a detailed pooled analysis of the dose-response relationship', *JAMA Internal Medicine*, vol. 175, pp. 959–67.

8 Hao S et al., (2016) 'Dietary Obesity Reversibly Induces Synaptic Stripping by Microglia and Impairs Hippocampal Plasticity', *Brain, Behavior, and Immunity*, vol. 51, pp. 230–39.

9 Hargrave SL et al., (2016) 'The Outward Spiral: a vicious cycle model of obesity and cognitive dysfunction', *Current Opinion in Behavioral Sciences*, vol. 9, pp. 40–46.

10 Alvarez-Crespo M et al., (2012) 'The Amygdala as a Neurobiological Target for Ghrelin in Rats: neuroanatomical, electrophysiological, and behavioral evidence', *PLoS One*, vol. 7, p. e46321.

11 Rendeiro C et al., (2015) 'The Mechanisms of Action of Flavonoids in the Brain: direct versus indirect effects', *Neurochemistry International*, vol. 89, pp. 126–39.

12 Stilling R et al., (2016) 'The Brain's Geppetto: microbes as puppeteers of neural function and behaviour?', *Journal of NeuroVirology*, vol. 22, pp. 14–21.

13 Klarer M et al., (2014) 'Gut Vagal Afferents Differentially Modulate Innate Anxiety and Learned Fear', *The Journal of Neuroscience*, vol. 34, pp. 7067–76.

14 Hebb D, (1949) *The Organization of Behavior*, Wiley & Sons, p. 62.

15 Woollett K & Maguire E, (2011) 'Acquiring "the Knowledge" of London's Layout Drives Structural Brain Changes', *Current Biology*, vol. 21, pp. 2109–14.

16 Draganski B et al., (2004) 'Neuroplasticity: changes in grey matter induced by training', *Nature*, vol. 427, pp. 311–12.

17 Zatorre RJ et al., (2012) 'Plasticity in Gray and White: neuroimaging changes in brain structure during learning', *Nature Neuroscience*, vol. 15, p. 528.

18 Zhang Y & Barres BA, (2013) 'A Smarter Mouse with Human Astrocytes', *Bioessays*, vol. 35, pp. 876–80.

19 Johansen-Berg H, (2007) 'Structural Plasticity: rewiring the brain', *Current Biology*, vol. 17, pp. R141–R144.

20 Melby-Lervåg M et al., (2016) 'Working Memory Training Does Not Improve Performance on Measures of Intelligence or Other Measures of "Far Transfer": evidence from a meta-analytic review', *Perspectives on Psychological Science*, vol. 11, pp. 512–34.

21 Au J et al., (2015) 'Improving Fluid Intelligence with Training on Working Memory: a meta-analysis', *Psychonomic Bulletin & Review*,

vol. 22, pp. 366–77.

22 Ibid.

## Chapter 1 – The Taming of the Butterfly

1 Killingsworth MA & Gilbert DT, (2010) 'A Wandering Mind is an Unhappy Mind', *Science,* vol. 330, p. 932.

2 Esterman M et al., (2013) 'In the Zone or Zoning Out?: tracking behavioral and neural fluctuations during sustained attention', *Cerebral Cortex,* vol. 23, pp. 2712–23.

3 Creswell JD et al., (2016) 'Alterations in Resting-State Functional Connectivity Link Mindfulness Meditation with Reduced Interleukin-6: a randomized controlled trial', *Biological Psychiatry,* vol. 80, pp. 53–61.

## Chapter 2 – Anxious All Areas

1 Russ TC et al., (2012) 'Association between Psychological Distress and Mortality: individual participant pooled analysis of 10 prospective cohort studies', *BMJ,* vol. 345, p. e4933.

2 Fuhrmann D et al., (2015) 'Adolescence as a Sensitive Period of Brain Development', *Trends in Cognitive Sciences,* vol. 19, pp. 558–66.

3 Elaine Fox is Professor of Cognitive and Affective Psychology at Oxford University and author of *Rainy Brain, Sunny Brain: the new science of optimism and pessimism*: www.rainybrainsunnybrain.com

4 In China, Korea, and Japan, for example, the SS-version of the transporter gene is the most common: Noskova T et al., (2008) 'Ethnic Differences in the Serotonin Transporter Polymorphism (5-HTTLPR) in Several European Populations', *Progress in Neuro-Psychopharmacology and Biological Psychiatry,* vol. 32, pp. 1735–39.

5 Pezawas L et al., (2005) '5-HTTLPR Polymorphism Impacts Human Cingulate-Amygdala Interactions: a genetic susceptibility mechanism for depression', *Nature Neuroscience,* vol. 8, pp. 828–34.

6 LeDoux J, (2015) *Anxious: the modern mind in the age of anxiety,* Oneworld, pp. 105–06.

7 You can try the version Elaine Fox uses in her studies — visit the Baldwin Social Cognition Lab website: http://baldwinlab.mcgill.ca/labmaterials/materials_BBC.html

8 Koster EHW et al., (2006) 'Attention to Threat in Anxiety-prone Individuals: mechanisms underlying attentional bias', *Cognitive Therapy and Research*, vol. 30, pp. 635–43.

9 http://psc.dss.ucdavis.edu/faculty_sites/sommerb/sommerdemo/stantests/norms.htm

## Chapter 3 – Let the Creativity Flow

1 McCaffrey T, (2012) 'Innovation Relies on the Obscure: a key to overcoming the classic problem of functional fixedness', *Psychological Science*, vol. 23, pp. 215–18.

2 Kounios J & Beeman M, (2009) 'The Aha! Moment: the cognitive neuroscience of insight', *Current Directions in Psychological Science*, vol. 18, pp. 210–16.

3 McPherson M et al., (2016) 'Emotional Intent Modulates the Neural Substrates of Creativity: an fMRI study of emotionally targeted improvisation in jazz musicians', *Scientific Reports*, vol. 6, article 18460.

4 Baas M et al., (2008) 'A meta-analysis of 25 Years of Mood-creativity Research: hedonic tone, activation, or regulatory focus?', *Psychological Bulletin*, vol. 134, p. 779–806.

5 Chermahini SA & Hommel B, (2010) 'The (B)link between Creativity and Dopamine: spontaneous eye blink rates predict and dissociate divergent and convergent thinking', *Cognition*, vol. 115, pp. 458–65.

6 Bentivoglio AR et al., (1997) 'Analysis of Blink Rate Patterns in Normal Subjects', *Movement Disorders*, vol. 12, pp. 1028–34.

7 Doughty MJ & Naase T, (2006) 'Further Analysis of the Human Spontaneous Eye Blink Rate by a Cluster Analysis-based Approach to Categorize Individuals with "Normal" Versus "Frequent" Eye Blink Activity', *Eye Contact Lens*, vol. 32, pp. 294–99.

8 Colzato L et al., (2015) 'Food for Creativity: tyrosine promotes deep thinking', *Psychological Research*, vol. 79, pp. 709–14.

## Chapter 4 – Lost in Space

1 Silverman I et al., (2007) 'The Hunter-Gatherer Theory of Sex Differences in Spatial Abilities: data from 40 countries', *Archives of Sexual Behaviour*, vol. 36, p. 261.

2 Hausmann M et al., (2000) 'Sex Hormones Affect Spatial Abilities during the Menstrual Cycle', *Behavioural Neuroscience*, vol 114, pp. 1245–50.

3 http://spatiallearning.org/resource-info/Spatial_Ability_Tests/sbsod_scale.pdf

4 Hartley T & Harlow R, (2012) 'An Association between Human Hippocampal Volume and Topographical Memory in Healthy Young Adults', *Frontiers in Human Neuroscience*, vol. 6, p. 338.

5 http://www.pnas.org/content/97/8/4398.full.pdf

6 Bannerman D et al., (2014) 'Hippocampal Synaptic Plasticity, Spatial Memory, and Anxiety', *Nature Reviews Neuroscience*, vol. 15, pp. 181–92.

7 Burles F et al., (2014) 'Neuroticism and Self-evaluation Measures Are Related to the Ability to Form Cognitive Maps Critical for Spatial Orientation', *Behavioural Brain Research*, vol. 271, pp. 154–59.

8 Kühn S et al., (2014) 'Playing Super Mario Induces Structural Brain Plasticity: gray matter changes resulting from training with a commercial video game', *Molecular Psychiatry*, vol. 19, pp. 265–71.

9 Hart V et al., (2013) 'Dogs Are Sensitive to Small Variations of the Earth's Magnetic Field', *Frontiers in Zoology*, vol. 10, p. 80.

10 Epstein RA & Vass LK, (2013) 'Neural Systems for Landmark-based Wayfinding in Humans', *Philosophical Transactions of the Royal Society B*, vol. 369.

11 Try the tests of Giuseppe Iaria's group at: www.gettinglost.ca

## Chapter 5 – Mind-bending, Time-bending

1 Droit-Volet S & Wearden J, (2016) 'Passage of Time Judgments Are Not Duration Judgments: evidence from a study using experience sampling methodology', *Frontiers in Psychology*, vol. 7, p. 176.

2 Fairhall SL et al., (2014) 'Temporal Integration Windows for Naturalistic Visual Sequences', *PLoS One*, vol. 9, p. e102248.

3 Pomares FB et al., (2011) 'How a Clock Can Change Your Pain?: the illusion of duration and pain perception', *Pain*, vol. 152, pp. 230–34.

4 Conti R, (2001) 'Time Flies: investigating the connection between intrinsic motivation and time awareness', *Journal of Personality*, vol. 69, pp. 1–26.

## Chapter 6 – Number Sense Lost

1 http://www.theguardian.com/education/2016/mar/26/reckon-you-were-born-without-a-brain-for-maths-highly-unlikely

2 OECD, (2016) *Equations and Inequalities: making mathematics accessible to all.*

3 Sarkar A et al., (2014) 'Cognitive Enhancement or Cognitive Cost: trait-specific outcomes of brain stimulation in the case of mathematics anxiety', *The Journal of Neuroscience*, vol. 34, pp. 16605–10.

4 Rubinsten O et al., (2012) 'Exploring the Relationship between Math Anxiety and Gender through Implicit Measurement', *Frontiers in Human Neuroscience*, vol. 6, p. 279.

5 Looi CY et al., (2016) 'Combining Brain Stimulation and Video Game to Promote Long-term Transfer of Learning and Cognitive Enhancement', *Scientific Reports*, vol. 6, article 22003.

6 Popescu T et al., (2016) 'Transcranial Random Noise Stimulation Mitigates Increased Difficulty in an Arithmetic Learning Task', *Neuropsychologia*, vol. 81, pp. 255–64.

7 Santarnecchi E et al., (2016) 'Individual Differences and Specificity of Prefrontal Gamma Frequency-tACS on Fluid Intelligence Capabilities', *Cortex*, vol. 75, pp. 33–43.

8 Bechara A et al., (2000) 'Emotion, Decision Making, and the Orbitofrontal Cortex', *Cerebral Cortex*, vol. 10, pp. 295–307.

9 http://www.kent.ac.uk/careers/tests/sequences.htm

## Chapter 7 – My Brain on Override

1 Gu S et al., (2015) 'Controllability of Structural Brain Networks', *Nature Communications,* vol. 6, article 8414.

2 *New Scientist,* 26 August 2015.

3 Hofmann SG et al., (2009) 'The Upside of Being Socially Anxious: psychopathic attributes and social anxiety are negatively associated', *Journal of Clinical and Social Psychology,* vol. 28, pp. 714–27.

4 Wager T & Atlas L, (2015) 'The Neuroscience of Placebo Effects: connecting context, learning and health', *Nature Reviews Neuroscience,* vol. 16, pp. 403–18.

5 Boot WR et al., (2013) 'The Pervasive Problem with Placebos in Psychology: why active control groups are not sufficient to rule out placebo effects', *Perspectives on Psychological Science,* vol. 8, pp. 445–54.

6 Ibid.

## Chapter 8 – The Road from Here

1 http://puzzlebox.io/orbit/

2 Arns M et al., (2014) 'Evaluation of Neurofeedback in ADHD: the long and winding road', *Biological Psychology,* vol. 98, pp. 108–15.

3 Reiter K et al., (2016) 'Neurofeedback Treatment and Posttraumatic Stress Disorder: effectiveness of neurofeedback on posttraumatic stress disorder and the optimal choice of protocol', *Journal of Nervous and Mental Disease,* vol. 204, pp. 69–77.

4 Enriquez-Geppert S et al., (2014) 'Self-regulation of Frontal-midline Theta Facilitates Memory Updating and Mental Set Shifting', *Frontiers in Behavioural Neuroscience,* vol. 8, p. 420.

5 Ros T et al., (2013) 'Mind over Chatter: plastic up-regulation of the fMRI salience network directly after EEG neurofeedback', *Neuroimage,* vol. 65, pp. 324–35.

6 Ruiz S et al., (2014) 'Real-time fMRI Brain Computer Interfaces: self-regulation of single brain regions to networks', *Biological Psychology,* vol. 95, pp. 4–20.

7 Ibid.

8 Santarnecchi E et al., (2016) 'Individual Differences and Specificity of Prefrontal Gamma Frequency-tACS on Fluid Intelligence Capabilities', *Cortex*, vol. 75, pp. 33–43.